Pietro Mantovan

Piercesare Secchi

Complex Data Modeling and Computationally Intensive Statistical Methods

Contributions to Statistics

For further volumes:
http://www.springer.com/series/2912

Pietro Mantovan (Editor)

Piercesare Secchi (Editor)

Complex Data Modeling and Computationally Intensive Statistical Methods

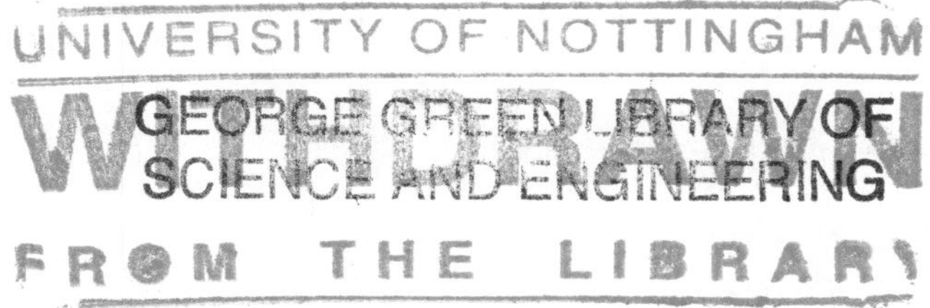

Physica-Verlag
A Springer Company

Pietro Mantovan
Ca' Foscari University of Venice
Venice, Italy

Piercesare Secchi
Politecnico di Milano
Milan, Italy

ISBN 978-88-470-1385-8 e-ISBN 978-88-470-1386-5

DOI 10.1007/978-88-470-1386-5

Library of Congress Control Number: 2010929480

Springer Milan Dordrecht Heidelberg London New York

Jointly published with Physica-Verlag Heidelberg, Germany

9 8 7 6 5 4 3 2 1

Cover-Design: WMX Design

Typesetting with LaTeX: PTP-Berlin, Protago TeX-Production GmbH, Germany (www.ptp-berlin.eu)
Printing and Binding: Grafiche Porpora, Segrate (Mi)
Printed in Italy

Springer-Verlag Italia srl – Via Decembrio 28 – 20137 Milano
Springer is a part of Springer Science+Business Media (www.springer.com)

Preface

Recent years have seen the advent and development of many devices able to record and store an ever increasing amount of information. The fast progress of these technologies is ubiquitous throughout all fields of science and applied contexts, ranging from medicine, biology and life sciences, to economics and industry. The data provided by these instruments have different forms: 2D-3D images generated by diagnostic medical scanners, computer vision or satellite remote sensing, microarray data and gene sets, integrated clinical and administrative data from public health databases, real time monitoring data of a bio-marker, system control datasets. All these data share the common characteristic of being complex and often highly dimensional.

The analysis of complex and highly dimensional data poses new challenges to the statistician and requires the development of novel models and techniques, fueling many fascinating and fast growing research areas of modern statistics. An incomplete list includes for example: functional data analysis, that deals with data having a functional nature, such as curves and surfaces; shape analysis of geometric forms, that relates to shape matching and shape recognition, applied to computational vision and medical imaging; data mining, that studies algorithms for the automatic extraction of information from data, eliciting rules and patterns out of massive datasets; risk analysis, for the evaluation of health, environmental, and engineering risks; graphical models, that allow problems involving large-scale models with millions of random variables linked in complex ways to be approached; reliability of complex systems, whose evaluation requires the use of many statistical and probabilistic tools; optimal design of computer simulations to replace expensive and time consuming physical experiments.

The contributions published in this volume are the result of a selection based on the presentations (about one hundred) given at the conference "*S.Co.2009*: Complex data modeling and computationally intensive methods for estimation and prediction", held at the Politecnico di Milano*. *S.Co.* is a forum for the discussion of new developments

* September 14–16, 2009. That of 2009 is its sixth edition, the first one being held in Venice in 1999.

and applications of statistical methods and computational techniques for complex and highly dimensional datasets.

The book is addressed to statisticians working at the forefront of the statistical analysis of complex and highly dimensional data and offers a wide variety of statistical models, computer intensive methods and applications.

We wish to thank all associate editors and referees for their valuable contributions that made this volume possible.

Milan and Venice, May 2010

Pietro Mantovan
Piercesare Secchi

Contents

List of Contributors

Alessandro Baldi Antognini
Department of Statistical Sciences
University of Bologna
Bologna, Italy

Graziano Aretusi
Department of Quantitative Methods and Economic Theory
University G. d'Annunzio
Chieti-Pescara, Italy

Raffaele Argiento
CNR IMATI
Milan, Italy

Pietro Barbieri
Ufficio Qualità
Cernusco sul Naviglio, Italy

Alessandro Barbiero
Department of Economics
Business and Statistics
University of Milan
Milan, Italy

Monica Chiogna
Department of Statistical Sciences
University of Padova
Padova, Italy

James M. Ciera
Department of Statistical Sciences
University of Padova
Padova, Italy

Thomas J. DiCiccio
Department of Social Statistics
Cornell University
Ithaca, USA

Lara Fontanella
Department of Quantitative Methods and Economic Theory
University G. d'Annunzio
Chieti-Pescara, Italy

Massimiliano Giorgio
Department of Aerospace and Mechanical Engineering
Second University of Naples
Aversa (CE), Italy

Niccolò Grieco
A.O. Niguarda Cà Granda
Milan, Italy

Maurizio Guida
Department of Electrical and Information Engineering
University of Salerno
Fisciano (SA), Italy

Alessandra Guglielmi
Department of Mathematics
Politecnico di Milano
Milan, Italy
also affiliated to CNR IMATI, Milano

Francesca Ieva
MOX – Department of Mathematics
Politecnico di Milano
Milan, Italy

Luigi Ippoliti
Department of Quantitative Methods and Economic Theory
University G. d'Annunzio
Chieti-Pescara, Italy

Sara Martino
Department of Mathematical Sciences
Norwegian University for Science and Technology
Trondheim, Norway

M. Sofia Massa
Department of Statistical Sciences
University of Padova
Padova, Italy

Fulvia Mecatti
Department of Statistics
University of Milano-Bicocca
Milan, Italy

Arcangelo Merla
Clinical Sciences and Bioimaging Department
Institute of Advanced Biomedical Technologies
Foundation University G. d'Annunzio
Chieti-Pescara, Italy

Anna Maria Paganoni
MOX – Department of Mathematics
Politecnico di Milano
Milan, Italy

Antonio Pievatolo
CNR IMATI
Milan, Italy

Gianpaolo Pulcini
Istituto Motori
National Research Council (CNR)
Naples, Italy

Chiara Romualdi
Department of Biology
University of Padova
Padova, Italy

Håvard Rue
Department of Mathematical Sciences
Norwegian University for Science and Technology
Trondheim, Norway

Piercesare Secchi
MOX – Department of Mathematics
Politecnico di Milano
Milan, Italy

Paolo Vidoni
Department of Statistics
University of Udine
Udine, Italy

G. Alastair Young
Department of Mathematics
Imperial College London
London, UK

Maroussa Zagoraiou
Department of Statistical Sciences
University of Bologna
Bologna, Italy

Enrico Zio
Ecole Centrale Paris-Supelec
Paris, France
and
Politecnico di Milano
Milan, Italy

Space-time texture analysis in thermal infrared imaging for classification of Raynaud's Phenomenon

Graziano Aretusi, Lara Fontanella, Luigi Ippoliti and Arcangelo Merla

Abstract. This paper proposes a supervised classification approach for the differential diagnosis of Raynaud's Phenomenon on the basis of functional infrared imaging (IR) data. The segmentation and registration of IR images are briefly discussed and two texture analysis techniques are introduced in a spatio-temporal framework to deal with the feature extraction problem. The classification of data from healthy subjects and from patients suffering from primary and secondary Raynaud's Phenomenon is performed by using Stepwise Linear Discriminant Analysis (LDA) on a large number of features extracted from the images. The results of the proposed methodology are shown and discussed for a temporal sequence of images related to 44 subjects.

Key words: Raynaud's Phenomenon, classification, functional infrared imaging, texture analysis, Gaussian Markov Random Fields

1 Introduction

Raynaud's Phenomenon (RP) is a paroxysmal vasospastic disorder of small arteries, pre-capillary arteries and cutaneous arteriovenous shunts of extremities, typically induced by cold exposure and emotional stress [2]. RP usually involves the fingers of the upper and lower extremities, even though tongue, nose, ears, and nipples may result affected as well. The presence of the initial ischemic phase is mandatory for clinical diagnosis, whereas reactive hyperaemic phase may not occur.

RP can be classified as primary (PRP), with no identifiable underlying pathological disorder, and secondary, usually associated with a connective tissue disease, the use of certain drugs, or the exposition to toxic agents [4]. Secondary RP is frequently associated with systemic sclerosis. In this case, RP typically may precede the onset of other symptoms and signs of disease by several years [2]. It has been estimated that 12.6% of patients suffering from primary RP develop a secondary disease. In particular, while between 5% and 20% of subjects suffering from secondary RP evolve in either limited or diffuse systemic sclerosis, all of the systemic sclerosis patients underwent or will experience RP [2]. These epidemiological data highlight the importance for early and proper differential diagnosis to distinguish the different forms of RP.

Mantovan, P., Secchi, P. (Eds.): Complex Data Modeling and Computationally Intensive Statistical Methods

Thermal infrared (IR) imaging has been widely used in medicine to evaluate cutaneous temperature. IR imaging is a non-invasive technique providing a map of the superficial temperature of a given body by measuring the emitted infrared energy [15]. Since the cutaneous temperature depends on local blood perfusion and thermal tissue properties, IR imaging provides important indirect information on circulation, thermal properties and thermoregulatory functionality of the cutaneous tissue. In this paper we thus exploit data from functional infrared imaging (fIRI) for classifying healthy controls (HCS), primary (PRP) and secondary (SSc) to systemic sclerosis RP patients.

The segmentation and registration of IR images are briefly discussed and two texture analysis techniques are introduced in a spatio-temporal framework to deal with the feature extraction problem. The classification is performed by using Stepwise Linear Discriminant Analysis (LDA) and the results of the proposed methodology are shown for a data set of 44 subjects.

The paper is organised as follows. In Section 2 we describe the data and how they have been created through a functional test; in Section 3 we deal with the processing of IR images by focusing on the problems of image segmentation and registration. Section 4 considers the problem of feature extraction and describes two different procedures for performing texture analysis in a spatio-temporal framework: one based on the estimation of a space-time Gaussian Markov Random Field, the other based on the calculation of texture measures obtained by co-occurrence matrices. Finally, in Section 5 we discuss classification results and in Section 6 we provide some conclusions.

2 The Data

Data for this study were provided by the Functional Infrared Imaging Lab – ITAB, Institute for Advanced Biomedical Technology, at the School of Medicine of the G. d'Annunzio University, Chieti, Italy. The study was approved by the local Institutional Review Boards and Ethics Committees. All subjects gave their informed written consent prior to being enrolled.

For each subject, raw data consist of a temporal sequence of 46 images, each of dimension (256×256), documenting the thermal recovery from a standardised cold stress produced in the hands of each subject [14]. Specifically, we have $n = 44$ subjects classified as follows: 13 HCS, 14 PRP, and 17 SSc. The classification was performed according to the American College of Rheumatology criteria and standard exclusion criteria were observed [15]. Furthermore, patients underwent thermal IR imaging after having observed standard preparatory rules to the test [14].

The thermal high-resolution IR images were acquired every 30 seconds to monitor the response to a cold stress. Images were acquired using a 14-bit digital thermal camera (FLIR SC3000 QWIP, Sweden) sensitive in the 8–9 μm band and with 0.02 K temperature resolution. To estimate the basal temperature of each subject, the image acquisition started 2.5 minutes before the cold stress and ended 20 minutes after. The cold stress consisted of a two minute immersion of the hands in cold water (at $10\,^{\circ}$C), while wearing thin plastic gloves. Since such a stress determines an immediate temperature drop from the stationary steady state, the study of the dynamics of the re-warming process is of particular interest. The recovery of the temperature is monitored starting from the sixth image of the series, that is just after the cessation of the cold stress.

3 Processing thermal high resolution infrared images

It is often a necessary step before a desired quantitative analysis to carry out a processing of the images. In particular, prior to the feature extraction for the classification of RP patients, we aim at constructing contours in the image to partition it into regions of interest (segmentation), and to perform spatial transformations with respect to an oriented reference image in order to compare images through the subjects (registration). Because of the inherent differences between infrared and visible phenomenology, a number of fundamental problems arise when trying to apply traditional processing methods to the infrared images [13, 17]. In the following we thus briefly consider the problems of segmentation and registration of IR images.

3.1 Segmentation

The nature of a thermal image is quite different from that of a conventional intensity image. In general, the latter encodes several physical properties such as reflectance, illumination and material of an object surface, to form the shape-related data, while a thermal image is formed by the heat distribution of an object[1].

Therefore, it is obvious that the conventional segmentation algorithms may not be feasible when they are applied to a thermal image [5, 10]. As usual, the purpose of thermal image segmentation is to separate objects of interest from the background, usually represented by thermal features showing a certain degree of spatial uniformity. In such cases, it would be possible to perform a segmentation by using a threshold procedure. However, due to the slight blurring caused by the infrared imaging process, it may happen that the boundary between a hand and the background is not so sharp for some of the images (see for example Figure 1).

In such cases, the images were segmented manually, pixel by pixel. An example of the results of the segmentation procedure is provided in Figure 2.

3.2 Registration

Image registration is the process of geometrical alignment of two images, a sensed image with respect to a reference image, required to obtain more complete and comparable information throughout the subjects. The majority of registration methods consists of the following steps [20]:

---

[1] Specifically, thermographic images depict the electromagnetic radiation of an object in the infrared range which is about 6–15 μm.

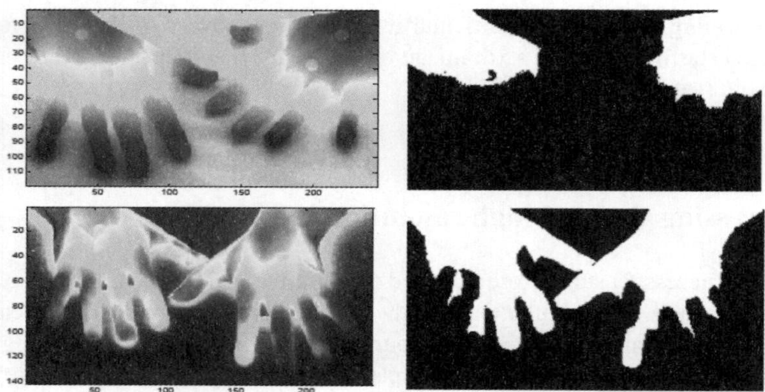

Fig. 1. Examples of IR image segmentation using threshold procedure

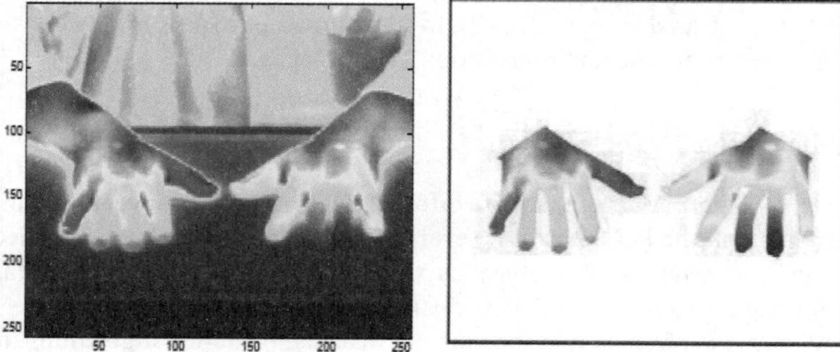

Fig. 2. Examples of manual IR image segmentation

- Manual or automated detection of distinctive objects represented by the so-called control points (i.e. input points on the sensed image and base points on the reference image);
- Estimation of the mapping function aligning the two images by matching the control points;
- Resample of the sensed image by means of the mapping function (image values in non-integer coordinates are computed by the appropriate interpolation method).

Due to the radically different ways of image formation in visible spectrum and thermographic images, many methods for registration of images work poorly or do not work at all [11]. A reasonable way to practice, is first to manually detect the control points, usually by using an aided procedure. Then, a set of mapping function parameters, valid for the entire image, are estimated to align the reference and the sensed images. In general, similarity transform, or affine transform, may be used in the mapping model; however, since in our study the distance, and the angle between the thermal camera and the scene are not always the same for all the subjects, a perspective projection model [20], with a bilinear interpolation method [18], was used to

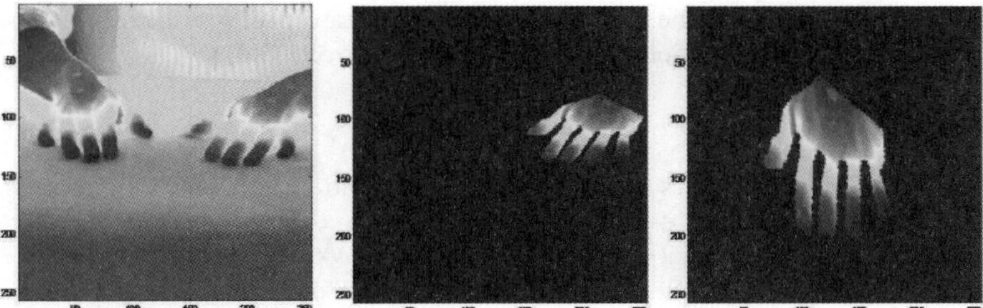

Fig. 3. A typical example of IR image: original image (left), segmented left-hand image (middle), registered image (right)

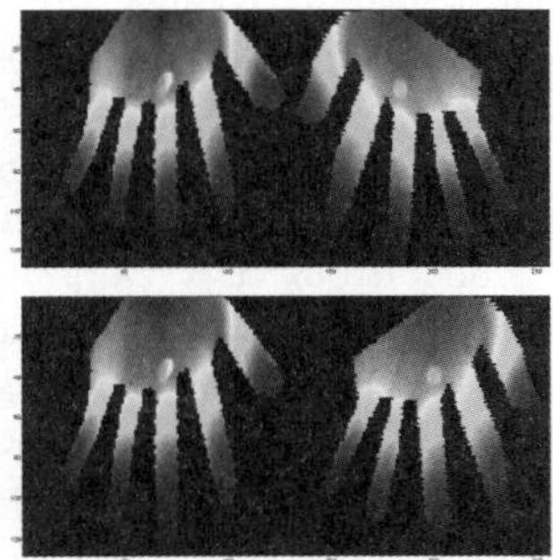

Fig. 4. Example of image reflection: registered image (top row), reflected image (bottom row)

perform the image registration. As an example, using 11 landmarks (control points), in Figure 3 we show the result of the registration for one of the segmented images.

Finally, in order to make the images spatially homogeneous, we also performed a reflection of the left hand with respect to its own longitudinal central axis. An example of image reflection is shown in Figure 4.

4 Feature extraction

With the aim of developing automatic discrimination techniques for HCS, PRP and SSc patients, we have to extract a set of features from the registered images. Such images display complex patterns at various orientations and we thus expect quite distinct texture characteristics among the classes. Texture analysis can be done either by studying the point properties of an image, in a pixel-based view, or explicitly defining the primitives that characterise the image, in a structural approach, to search for features such as spatial arrangement. In this section, we describe in detail two procedures. In the first, the temperature values are considered as a realisation of a spatio-temporal Gaussian Markov Random Field (ST-GMRF) such that the parameters of such a process represent part of the full set of variables to be used in the classification process. In the second procedure, other features of interest are obtained by extracting the information from co-occurrence matrices (CMs).

4.1 ST-GMRFs

Let the temperature values of the hands be a realisation of a stochastic process, $\mathbf{X} = \{X(\mathbf{p}), \mathbf{p} = (\mathbf{s}, t) \in \Delta \subset \Re^d\}$, defined on a d-dimensional space, where $d = d_s + d_t$ with d_s spatial dimensions and d_t the temporal dimension. For $d_s = 2$ and $d_t = 1$, $\{X(\mathbf{s}, t), \mathbf{s} \in \Delta_S \subset \Re^2, t \in \Delta_T \subset \Re^+\}$ is a spatio-temporal random field (ST-RF).

Suppose that the spatio-temporal process $\mathbf{X} = \{X(\mathbf{p}), \mathbf{p} \in \Delta \subset \Re^3\}$ has mean $\mu(\mathbf{p})$ representing a parameterised unknown deterministic spatio-temporal trend function. We will assume that the residual process $\varepsilon(\mathbf{p}) = X(\mathbf{p}) - \mu(\mathbf{p})$ is a zero mean stationary Gaussian process, with covariance function given by $\sigma(\mathbf{h}, k) = Cov[X(\mathbf{s}, t), X(\mathbf{s} + \mathbf{h}, t + k)]$, where $\mathbf{h} = \mathbf{s}_i - \mathbf{s}_j$ and $k = t_i - t_j$. If $\sigma(\mathbf{h}, k) = \sigma(||\mathbf{h}||, k)$, where $||\mathbf{h}|| = \sqrt{\mathbf{h}'\mathbf{h}}$, the ST-RF is said to be spatially isotropic.

A spatio-temporal Random Field (ST-RF) X observed on a spatio-temporal lattice $L \subset \Delta$, with g grid points, is denoted by $\mathbf{X} = [X(\mathbf{p}_1), X(\mathbf{p}_2), \ldots, X(\mathbf{p}_g)]'$ and is characterised by a $(g \times 1)$ mean vector, μ, and a $(g \times g)$ covariance matrix, Σ. Notice that for a 3-dimensional regular lattice, we have $g = N \times M \times T$.

The mean structure can be modeled through a linear combination of independent variables with unknown parameters $\mathbf{b}$; i.e. $\boldsymbol{\mu} = \mathbf{Db}$; specifically, if we consider a spatio-temporal trend-surface analysis, the entries of the design matrix $\mathbf{D}$ are expressed as a function of the coordinates of site $\mathbf{p} = (\mathbf{s}, t)$.

Dealing with huge data sets, as in our case, it may be better for computational purposes to assume a conditional specification of the process such that, under Gaussian assumptions, $\mathbf{X} \sim N(\mathbf{Db}, \Sigma)$, where $\Sigma = v^2\mathbf{A}^{-1}$. v^2 is the conditional variance while $\mathbf{A}$ is the *potential matrix* with entries equal to 1 along the main diagonal, the inverse correlations $-c_{ij}$ if the sites $\mathbf{p}_i$ and $\mathbf{p}_j$ are neighbours, and otherwise zero.

Therefore, $\mathbf{X}$ is a Space-Time Gaussian Markov Random Field (ST-GMRF) with conditional mean and conditional variance given by

$$E\left(X(\mathbf{p}_i)|X(\mathbf{p}_j), j \neq i\right) = \mu(\mathbf{p}_i) + \sum_{j \neq i} c_{ij}\left[X(\mathbf{p}_j) - \mu(\mathbf{p}_j)\right]$$

$$Var\left(X(\mathbf{p}_i)|X(\mathbf{p}_j), j \neq i\right) = v^2,$$

where c_{ij} are the spatio-temporal interaction parameters.

For a homogeneous process, and for a displacement vector $\mathbf{r}$, the conditional mean can also be rewritten as

$$E\left(X(\mathbf{p}_i)|X(\mathbf{p}_j), j \neq i\right) = \mu(\mathbf{p}_i) + \sum_{\mathbf{r} \in \delta_i^\alpha} c_{\mathbf{r}}\left[X(\mathbf{p}_i + \mathbf{r}) - \mu(\mathbf{p}_i + \mathbf{r})\right], \quad (1)$$

where δ_i^α is the set of neighbours of pixel $\mathbf{p}_i$, α is the ST-GMRF *order* which is defined on a rectangular lattice by a given maximum distance between two pixels, $c_{\mathbf{r}} = c_{-\mathbf{r}}$, $c_{\mathbf{0}} = 0$, and $c_{\mathbf{r}} = 0, \forall \mathbf{r} : \mathbf{p}_i + \mathbf{r} \notin \delta_i^\alpha$.

Considering a pure spatial process, $\mathbf{r}$ is a 2-dimensional vector, and for a first order ($\alpha = 1$) homogeneous GMRF we have two spatial interaction parameters for $\mathbf{r} = (0, 1)$ and $\mathbf{r} = (1, 0)$ for neighbours which are one pixel apart horizontally and vertically, respectively. For a second order GMRF we have four interaction parameters for $\mathbf{r} = (0, 1)$ and $\mathbf{r} = (1, 0)$ together with $\mathbf{r} = (1, 1)$ and $\mathbf{r} = (1, -1)$ for diagonally adjacent neighbours in the South-East and North-West directions, respectively.

The neighbourhood structure of a ST-GMRF can be defined by specifying the space-time neighbourhood from a time series perspective. In this case the order is provided by a vector $\boldsymbol{\alpha} = (\alpha_s, \alpha_t)$ consisting of the spatial and temporal lags, respectively [9]. In general, we will have $(1 + 2\alpha_1)(1 + \alpha_2)$ parameters for a homogeneous process, and $(1 + \alpha_1)(1 + \alpha_2)$ parameters for a completely symmetric one.

An important part of the ST-GMRF model specification is the choice of boundary conditions (b.c.) for a stationary process, since elements of Σ^{-1} for boundary sites on a finite lattice can be very complicated [3]. In general, to deal with ST-GMRFs on finite rectangular lattices, the most convenient boundary conditions are toroidal b.c. These specify that each dimension is assumed to be wrapped around, so that the first and last coordinates are adjacent. There are many possible methods for estimating the parameter vector $\boldsymbol{\eta} = (\mathbf{b}', \mathbf{c}', \upsilon^2)'$. ML minimises the negative log-likelihood

$$L(\mathbf{b}, \mathbf{c}, \upsilon^2) = \frac{g}{2} log(2\pi\ \upsilon^2) - \frac{1}{2} log(|\mathbf{A}|) + \frac{1}{2\upsilon^2}(\mathbf{x} - \mathbf{Db})'\mathbf{A}(\mathbf{x} - \mathbf{Db}), \qquad (2)$$

where $\mathbf{A}$ is symmetric and positive-definite. The negative log-likelihood can be minimised in stages [6]. In fact,

- conditional on $\mathbf{A}$

$$\hat{\mathbf{b}} = (\mathbf{D}'\mathbf{AD})^{-1}\mathbf{D}'\mathbf{Ax} \qquad (3)$$

$$\hat{\upsilon}^2 = \frac{1}{g}(\mathbf{x} - \mathbf{Db})'\mathbf{A}(\mathbf{x} - \mathbf{Db}). \qquad (4)$$

  Equations (3) and (4) provide the m.l. estimators of the trend parameters $\mathbf{b}$ and the conditional variance υ^2.

- Then, substituting (3) and (4) back into Equation (2), the m.l. estimators of the spatial interaction parameters $\mathbf{c}$ can be obtained by minimising the *profile likelihood*

$$L^*(\mathbf{c}|\hat{\upsilon}^2, \hat{\mathbf{b}}) \propto -\frac{1}{2} log(|\mathbf{A}|) + \frac{g}{2} log\left[\mathbf{x}'\mathbf{A}\left\{\mathbf{I} - \mathbf{QA}\right\}\mathbf{x}\right] \qquad (5)$$

  where $\mathbf{Q} = \mathbf{D}(\mathbf{D}'\mathbf{AD})^{-1}\mathbf{D}'$.
  When toroidal b.c. are assumed, minimisation of (5) for parameter estimation can be carried out with only O (g log g) steps [3, 8]; in fact, since the $(g \times g)$ spatial interaction matrix $\mathbf{A}$ is block circulant, and each block is itself block-circulant, the likelihood is that (5) can be evaluated at a low computational cost by means of a 3-dimensional discrete Fourier transform. For a discussion on the advantages of using toroidal b.c. see, for example, [8, 16 and 9].

4.2 Texture statistics through co-occurrence matrices

In this section, we use a pixel-based approach to identify further basic patterns that could represent the natural texture structure of the RP. Specifically, we perform here a texture analysis by extracting information in the form of a co-occurrence matrix (CM) and by summarising this information through the calculation of some measures of texture on the CM [7].

To calculate these measures, at each time t, we first classify the estimated residual process $\hat{\varepsilon}(\mathbf{p})$ in L levels, where L, is chosen by considering the quantiles of the space-time temperature distribution of all the subjects. Then, for a given space-time displacement vector, $\mathbf{r}$, which defines pairs of neighbours in the temporal, spatial and spatio-temporal domain, we compute the CM which provides a tabulation of how often different combinations of classified pixel values occur in an image [7]. More specifically, the $(i, j)th$ element of the $(L \times L)$ CM, denoted here as $\mathbf{C_r}$, represents the relative frequency, $f(i, j)$, of occurrence of a pair of classified pixel values, separated by the displacement $\mathbf{r}$ and having temperature levels i and j, respectively. Therefore, for a displacement vector $\mathbf{r}$, we calculate the set of the following texture measures:

$$T_1(\mathbf{r}) = \sum_{i,j}(i-j)^2 f(i,j), \quad T_2(\mathbf{r}) = \sum_{i,j}\frac{f(i,j)}{1+|i-j|}, \quad T_3(\mathbf{r}) = \sum_{i,j} f(i,j)^2,$$

$$T_4(\mathbf{r}) = \frac{\sum_{i,j} i\, j f(i,j) - \sum_i i f(i,\cdot)\sum_j j f(\cdot,j)}{\sigma_i\, \sigma_i},$$

$$T_5(\mathbf{r}) = \sum_{i,j} f(i,j)\log_2 \frac{f(i,j)}{f(i,\cdot)f(\cdot,j)},$$

where

$$\sigma_i = \left[\sum_i i^2 f(i,\cdot) - \left(\sum_i i f(i,\cdot)\right)^2\right]^{1/2},$$

$$\sigma_j = \left[\sum_j j^2 f(\cdot,j) - \left(\sum_j j f(\cdot,j)\right)^2\right]^{1/2}$$

and $f(\cdot, j)$ and $f(i, \cdot)$ represent the marginal frequencies over the indices j and i, respectively.

The indices T_1 and T_2 represent *Contrast* and *Homogeneity* measures and use weights related to the distance from the diagonal of the CM; T_3 is known as *Energy* and gives information about orderliness; finally, T_4 and T_5 are *Correlation* and *Mutual Information* indices, respectively; they provide a measure of the linear and non-linear dependence of pairs of classified pixel values.

5 Classification results

In this section we discuss discrimination results on the data described in Section 2. For each subject, we have a spatio-temporal data matrix representing a temporal sequence of registered images. For computational purposes we perform the analysis of the segmented left and right hand images separately. In total, we have $n = 44$ subjects classified as HCS, PRP, and SSc. Since the re-warming process changes slowly, we have downsampled the time series by considering one image per minute; for each subject, and for each hand, we thus have a sequence of 19 images, each of dimension (128×128).

The identification of the feature variables for each subject, starts with the estimation of the parameters of a ST-GMRF which is very commonly used for modeling textures in image analysis [8]. The estimated mean function is based on a spatio-temporal function expressed as a polynomial function of time and spatial coordinates; more specifically, we consider a trend which is linear in space and quadratic in time; interaction terms between space and time are also included in the model. Overall, the estimated trend function is represented by 7 parameters. For the residual correlated process, we consider a type-ST neighbourhood structure with 4 neighbours in space ($\alpha_1 = 2$) and one lag in time ($\alpha_2 = 1$); this corresponds to a ST-GMRF with 10 parameters to be estimated. However, notice that considering both hands, this procedure leads to the estimation of 34 parameters for the whole image.

As regards the use of co-occurrence matrices, we consider ten levels ($L = 10$) and spatial displacements corresponding to the four main spatial directions (i.e., East, West, North, South). Considering a spatial lag up to 6, all the displacements can be collected in a global vector, $\mathbf{d}$, which takes the following structure: $\mathbf{d} = [(0\ 1); (0\ 2); \ldots; (0\ 6); (-1\ 1); (-2\ 2); \ldots; (-6\ 6); (-1\ 0); (-2\ 0); \ldots; (-6\ 0); (-1\ -1); (-2\ -2); \ldots; (-6\ -6)]$. Thus, overall we specify 24 different spatial lags and, for each of them, we can calculate 19 CMs. The joint use of a first order temporal lag then completes the specification of the temporal and spatio-temporal neighbourhood structure. Therefore, both for the first order temporal lag and for each of the 24 spatio-temporal displacements, we can compute further 18 CMs. However, to avoid an increase of the number of discriminant variables, for each temporal, spatial and spatio-temporal displacement, $\mathbf{r}$, we aggregate the frequencies of both hands corresponding to each pair of levels (i, j), thus obtaining a synthesised CM matrix, $\tilde{\mathbf{C}}_{\mathbf{r}}$, from which we can calculate the five texture measures, $T_1, \cdots, T_5$. This procedure, generates 245 variables; by adding them to the ones provided by the specification of the ST-GMRF, for each subject we thus have a total of 279 variables.

Of course, it is highly likely that a large number of these features do not provide any significant discriminatory information. Furthermore, the classification based on such a large number of variables may tend to be overfitting. Hence, to reduce the number of variables to a suitable number for the classification routine we use a forward stepwise linear discriminant analysis based on the Mahalanobis distance [12].

The best subset of selected features consists of a total of 14 discriminant variables, mainly related to the vertical and diagonal directions. Specifically, six of the selected variables are related to the parameters of the ST-GMRF while the remaining ones are represented by the indices T_1, T_4 and T_5. Variables related to the diagonal directions are characterised by a displacement of maximum 2 lags while variables related to the vertical directions are defined for spatial lags ranging from 4 to 6. The wider spatial lags observed for the vertical direction could be likely linked to the specific geometry of the finger vasculature and the expression of the functional impairment secondary to the disease. In fact, larger finger vessels run longitudinally and parallelously, while only a few arteriovenous shunts run transversally. Moreover, the presence of scleroderma leads to a progressive destruction of the microvasculature from distal to proximal sections, thus explaining the differences observed in the spatial lags.

The confusion matrix resulting from LDA gives a 0% estimate of the apparent error rate, but performing the Leave-One-Out Cross Validation (LOCV) procedure, the estimated error rate increases up to 9.1%. Specifically, Table 1 shows the confusion matrix for the LOCV procedure where it can be seen that, 2 HCS, 1 PRP and 1 SSc are wrongly classified.

Table 1. Confusion matrix for the LOCV procedure

	Predicted Groups			
Original	HCS	PRP	SSc	Total
HCS	11	0	2	13
PRP	0	13	1	14
SSc	1	0	16	17

6 Conclusions

In this paper we have discussed the problem of classifying functional infrared imaging data for differential diagnosis of primary and secondary RP forms. The segmentation and registration of IR images were also briefly considered. Specifically, we have noticed that segmentation of IR images is not an easy task. It is true that the infrared image acquisition is a part of the problem in the infrared segmentation process. In fact, the technologies required for IR imaging are much less mature than the ones used in visible imaging [17]. However, nowadays, cameras that collect simultaneously both visual and thermal images are available, and performing a fusion of the two types of images may lead to an improvement of the results.

Classification of primary and secondary RP forms has been performed within a spatio-temporal context using texture analysis techniques. We have proposed two different techniques which were able to generate a large number of discriminant variables. Good classification results, characterised by an error rate of 9.1%, were achieved through LDA; to our knowledge there are no other works which achieve such results for Raynaud's Phenomenon.

Due to the characteristics of the generated data set, of primary importance are of course those classifiers that can provide a rapid analysis of large amounts of data and have the ability to handle *fat data*, where there are more variables than observations. In this paper we considered forward stepwise LDA, but other approaches, based for example on Discriminant Partial Least Square [1] or shrinkage methods (eg. Lasso) [19], will be of particular interest for future works.

As a final consideration we do recognise that the small sample size of the study could be an obstacle for a strong interpretation of the statistical results. Unfortunately, the small sample size is something that cannot be changed in practical problems like this, but we aim at extending the analysis to a wider data set (presently being acquired) in further works.

References

1. Barker, M., Rayens, W.: Partial Least Squares for Discrimination. Journal of Chemometrics **17**, 166–173 (2003)
2. Belch, J.: Raynaud's phenomenon. Its relevance to scleroderma. Ann. Rheum. Dis. **50**, 839–845 (2005)
3. Besag, J.E., Moran, A.P.: On the estimation and testing of spatial interaction in Gaussian lattice processes. Biometrika **62**, 555–562 (1975)
4. Block, J.A., Sequeira, W.: Raynaud's phenomenon. Lancet **357**, 2042–2048 (2001)
5. Chang, J.S., Liao, H.Y.M., Hor, M.K., Hsieh, J.W., Cgern, M.Y.: New automatic multilevel thresholding technique for segmentation of thermal images. Images and vision computing **15**, 23–34 (1997)
6. Cressie, N.A.: Statistics for spatial data. second edn., Wiley and Sons, New York, (1993)
7. Cocquerez, J.P., Philipp, S.: Analyse d'images: filtrage et segmentation. Masson, Paris (1995)
8. Dryden, I.L., Ippoliti, L., Romagnoli, L.: Adjusted maximum likelihood and pseudolikelihood estimation for noisy Gaussian Markov random fields. Journal of Computational and Graphical Statistics **11**, 370–388 (2002)
9. Fontanella, L., Ippoliti, L., Martin, R.J., Trivisonno, S.: Interpolation of Spatial and Spatio-Temporal Gaussian Fields using Gaussian Markov Random Fields. Advances in Data Analysis and Classification **2**, 63–79 (2008)
10. Heriansyak, R., Abu-Bakar, S.A.R.: Defect detection in thermal image for nondestructive evaluation of petrochemical equipments. NDT&E International **42**, 729–740 (2009)
11. Jarc, A., Pers, J., Rogelj, P., Perse, M., Kovacic, S.: Texture features for affine registration of thermal and visible images. Computer Vision Winter Workshop (2007)
12. Johnson, R.A., Wichern, D.W.: Applied Multivariate Statistical Analysis. sixth edn., Pearson education, Prentice Hall, London (2007)
13. Maldague, X.P.V.: Theory and practice of infrared technology for nondestructive Testing. Wiley Interscience, New York (2001)
14. Merla, A., Romani, G.L., Di Luzio, S., Di Donato, L., Farina, G., Proietti, M., Pisarri, S., Salsano, S.: Raynaud's phenomenon: infrared functional imaging applied to diagnosis and drug effect. Int. J. Immunopathol. Pharmacol. **15**(1), 41–52 (2002a)
15. Merla, A., Di Donato, L., Pisarri, S., Proietti, M., Salsano, F., Romani, G.L.: Infrared Functional Imaging Applied to Raynaud's Phenomenon. IEEE Eng. Med. Biol. Mag. **6**(73), 41–52 (2002b)
16. Rue, H., Held, L.: Gaussian Markov random Fields. Theory and Applications. Chapman and Hall/CRC, Boca Raton (2005)

17. Scribner, D.A., Schuller, J.M., Warren, P., Howard, J.G., Kruer, M.R.: Image preprocessing for the infrared. *Proceedings of SPIE*, the International Society for Optical Engineering **4028**, 222–233 (2000)
18. Semmlow, J.L.: Biosignal and Biomedical Image Processing. CRC Press, Boca Raton (2004)
19. Tibshirani, B.: Regression shrinkage and selection via the Lasso. Journal of the Royal Statistical Society, Series B, **58**, 267–288 (1996)
20. Zitova, B., Flusser, J.: Image registration methods: a survey. Image and Vision Computing **21**, 977–1000 (2003)

Mixed-effects modelling of Kevlar fibre failure times through Bayesian non-parametrics

Raffaele Argiento, Alessandra Guglielmi and Antonio Pievatolo

Abstract. We examine the accelerated failure time model for univariate data with right censoring, with application to failure times of Kevlar fibres grouped by spool, subject to different stress levels. We propose a semi-parametric modelling by letting the error distribution be a shape-scale mixture of Weibull densities, the mixing measure being a normalised generalised gamma measure. We obtain posterior estimates of the regression parameter and also of credibility intervals for the predictive distributions and their quantiles, by including the posterior distribution of the random mixing probability in the MCMC scheme. The number of components in the non-parametric mixture can be interpreted as the number of groups, having a prior distribution induced by the non-parametric model, and is inferred from the data. Compared to previous results, we obtain narrower interval estimates of the quantiles of the predictive survival function. Other diagnostic plots, such as predictive tail probabilities and Bayesian residuals, show a good agreement between the model and the data.

Key words: accelerated failure time regression models, Bayesian semiparametrics, MCMC algorithms, mixed-effects models, mixture models, Weibull distribution

1 Introduction

We will present a Bayesian semiparametric approach for an Accelerated Failure Time (AFT) model for censored univariate data, on the basis of an application involving pressure vessels, which are critical components of the Space Shuttle. Given a fixed p-vector of covariates $x = (x_1, \ldots, x_p)'$ the failure time T is modelled as $\log T = x'\beta + W$, where $\beta = (\beta_1, \ldots, \beta_p)'$ is the vector of regression parameters and W denotes the error. Recently this model has received much attention in the Bayesian community, in particular in papers where the error W (or $\exp(W)$) has been represented hierarchically as a mixture of parametric densities with a Dirichlet process as mixing measure, *i.e.*, the well-known Dirichlet process mixture (DPM) models, introduced by Lo [16]. Kuo and Mallick [13] were the first to model $\exp(W)$ as a Dirichlet process location mixture, in particular assuming a DPM of normals with fixed and small variance. Kottas and Gelfand [12] and Gelfand and Kottas [6] propose a flexible semiparametric class of zero median distributions for the error,

Mantovan, P., Secchi, P. (Eds.): Complex Data Modeling and Computationally Intensive Statistical Methods

which essentially consists of a Dirichlet process scale mixture of split 0-mean normals (with skewness handled parametrically); this model seems very useful for estimating regression effects and for survival analysis where it is known a priori that the error distribution is unimodal. Since the marginal prior for $\exp(W)$ in Kuo and Mallick [13] gives positive probability to the negative reals, Hanson [8] proposes a DPM model of gamma densities, mixing over both the scale and the shape of the gammas, for the distribution of $\exp(W)$. Argiento *et al.* [1] compare models in the same setting when the mixing measure is either a Dirichlet process or a normalised inverse Gaussian process.

The application considered here is based on a dataset of 108 lifetimes of pressure vessels wrapped with a commercial fibre called Kevlar, obtained from a series of accelerated life tests; the fibre comes from eight different spools and four levels of stress (pressure) are used. Eleven lifetimes with the lowest level of stress are administratively censored at 41,000 hours. Crowder *et al.* [5] present a frequentist analysis of these data and fit an AFT model with both stress and spool as fixed effects, using Weibull distributions to model survival times. León *et al.* [14] take a Bayesian parametric approach to the problem, still using Weibull survival times, however considering a mixed-effects model, where the parameters for the spool effect in the linear predictor are exchangeable. This choice follows from the finding (common to all the cited references) that spools have a significant effect on the failure time, so that it is necessary to have a model for predicting the failure time when a new spool is selected at random from the population of spools. As a further generalisation, a greater degree of model flexibility can be obtained by imposing a non-parametric hierarchical mixture on the error term. We will see that a consequence of this assumption is a new representation of the exchangeable spool effect parameters in the model as mixture components. We then consider $T = \exp(x'\beta)\,V,\;\; V := \mathrm{e}^W$ where the error distribution is represented as a non-parametric hierarchical mixture of Weibull distributions on both the shape and the rate parameters. One of the distinctive features of our mixtures with respect to the previously proposed Bayesian semiparametric models is the use of a normalised generalised gamma measure as a prior for the mixing distribution G. Such a family is more flexible that Dirichlet (which is included here), since it is indexed by an additional parameter which controls the clustering property of the non-parametric hierarchical mixture. The Bayesian semiparametric approach makes it also possible to obtain interval estimates for the predictive distribution of the failure time of a vessel under a given stress condition and for its quantiles: a characterisation of G in terms of Poisson processes allows in fact its direct MCMC simulation. With respect to the Bayesian mixed-effects model, another feature of our approach is that the grouping of observations is not fixed (as dictated by the spool number), but is random and is inferred from the data. This information however is not lost, but is included in the prior distribution via the hyperparameters. Therefore, membership to a group is modelled non-parametrically, thanks to the clustering property of discrete random probability measures such as G. Other papers assuming random effects modelled non-parametrically include Kleinman and Ibrahim [11] and Jara *et al.* [10].

Compared to previous analyses, the non-parametric mixture model for the error better follows the log-linear relationship between failure times and covariates. Other diagnostic plots, such as the predictive tail probabilities and the Bayesian residuals show good agreement between the model and the data. Finally, the interval estimates obtained with the non-parametric shape-rate mixture error model are narrower than those under the parametric mixed-effects model.

The paper is organised as follows. Section 2 presents the AFT model used so far to analyse the dataset considered. In Section 3 we introduce our semiparametric AFT model and give the structure of the non-parametric hierarchical mixture prior to the error term. The application to Kevlar fibres is presented in Section 4, while conclusions and comments are given in Section 5. The Appendix sketches the algorithm used to compute inferences.

2 Accelerated life models for Kevlar fibre life data

Crowder *et al.* [5] consider 108 Kevlar fibre lifetimes, coming from the combination of eight different spools and four levels of stress (pressure), and fit an AFT model with both stress and spool as fixed effects, assuming a Weibull distribution for the lifetimes. The model can be described as follows: for stress level x, spool i and test number j,

$$T_{xij} \overset{ind}{\sim} \text{Weibull}(r, \lambda_{xi}), \quad j = 1, \ldots, n_{xi}, \; i = 1, \ldots, 8, \tag{1}$$

where n_{xi} denotes the number of tests from spool i and stress level x and $r > 0$, or equivalently, the distribution function of T_{xij} from spool i is

$$F_{xi}(t) = 1 - \mathrm{e}^{-\left(\frac{t}{\lambda_{xi}}\right)^r}, \quad t > 0, \tag{2}$$

with

$$\log(\lambda_{xi}) = \beta_0 + \beta_1 \log x + \alpha_i, \quad i = 1, \ldots, 8. \tag{3}$$

Crowder *et al.* [5] assume the log-stress as well as the spool as fixed effects. They find that the spool effect is very significant in the model and obtain an acceptable fit as far as the plot of residuals are concerned, but with a less satisfactory performance for the lowest stress level. In a recent paper, León *et al.* [14] argue that the fixed-effects model does not allow to make inference on vessels wrapped with fibre taken from a new spool. For the likelihood as in (1), they fit a Bayesian mixed-effects model and also a Bayesian fixed-effects model, the latter mainly for comparison with the frequentist model. More specifically, León *et al.* [14] assume the spool as a random effect, that is, $(\alpha_1, \ldots, \alpha_8)$ in (2)-(3) are conditionally i.i.d. $N(0, \sigma_\alpha^2)$ with an inverse gamma prior for σ_α^2, and β_0 and β_1 are given independent priors, as customary in Bayesian regression models. Then, it becomes possible to evaluate the predictive distribution of the life T_{109} of a new pressure vessel wrapped with fibre from a ninth spool, subject to a log-stress x:

$$f(t_{109} \mid \text{data}, x) = \int f(t_{109} \mid \beta, \alpha_9, r, x) f(\beta, \alpha_9, r \mid \text{data})\, d\beta d\alpha_9 d r, \quad \text{where}$$

$$f(\beta, \alpha_9, r \mid \text{data}) = \int f(\alpha_9 \mid \sigma_\alpha^2) f(\beta, r, \sigma_\alpha^2 \mid \text{data})\, d\sigma_\alpha^2 .$$

For every tested spool and for a new spool, the authors calculate point and interval estimates for the first percentile of the failure time distribution when stress is 23.4 MPa (MegaPascal), the lowest value in the dataset, and for the median with stress 22.5 MPa. With regards to the eight tested spools, they observe that the Bayesian mixed-effects model produces a shrinkage effect of the estimates for the more extreme spools toward a middle value, compared to the fixed-effects model; the interval estimates are also narrower on average. With regards to a new spool, the resulting prediction intervals are too wide to make statements about the reliability of the Space Shuttle.

3 The Bayesian semiparametric AFT model

In order to describe the Bayesian semiparametric model, it is better to rewrite model (1) as

$$T_{xij} = \exp(\beta_1 \log x + \alpha_i) V_{ij}, \quad j = 1, \ldots, n_{xi}, \ i = 1, \ldots, 8,$$

where

$$V_{ij} \overset{iid}{\sim} \text{Weibull}(\vartheta_1, \vartheta_2), \quad \vartheta_2 = \mathrm{e}^{\beta_0}.$$

To assume a more flexible model, we replace the Weibull distribution for the error with a non-parametric mixture of Weibulls, so that:

$$\begin{aligned} &T_{xij} = \exp(\beta_1 \log x + \alpha_i) V_{ij}, \quad j = 1, \ldots, n_{xi}, \ i = 1, \ldots, 8, \\ &V_{ij} \mid \vartheta_{1ij}, \vartheta_{2ij} \overset{ind}{\sim} \text{Weibull}(\vartheta_{1ij}, \vartheta_{2ij}) \\ &(\vartheta_{1ij}, \vartheta_{2ij}) | G \overset{iid}{\sim} G, \\ &G \sim NGG(\sigma, \eta, G_0) \\ &\beta_1 \sim \pi, \quad \beta_1 \perp G. \end{aligned} \tag{4}$$

Observe that, under this assumption, given G, the V_{ij} are independent and identically distributed according to the (non-parametric mixture) distribution function

$$F(v; G) = \int_\Theta \left(1 - \mathrm{e}^{-\left(\frac{v}{\vartheta_2}\right)^{\vartheta_1}}\right) G(d\vartheta_1, d\vartheta_2). \tag{5}$$

The mixing distribution assumed for the error V is a (a.s.) discrete random measure G, namely G is a normalised generalised gamma process, in short $G \sim NGG(\sigma, \eta, G_0)$. The usual choice for V is a Dirichlet process mixture, that is, the non-parametric process prior for G is a Dirichlet process. However, the NGG process prior, which is now briefly described, is more general and includes the Dirichlet process. Let Θ be a Borel subset of $\mathbb{R}^s$ for some positive integer s, with its Borel σ-algebra $\mathcal{B}(\Theta)$, and let G_0 be a probability on Θ. G is a normalised generalised gamma random probability measure on Θ when G can be written as

$$G = \sum_{i=1}^{+\infty} P_i \delta_{X_i} = \sum_{i=1}^{+\infty} \frac{J_i}{T} \delta_{X_i}, \tag{6}$$

where $P_i := \frac{J_i}{T}$ for any $i = 1, 2, \ldots, T := \sum_i J_i$, and $J_1 \geq J_2 \geq \ldots$ are the ranked values of points in a Poisson process on $(0, +\infty)$ with intensity

$$\varrho(ds) = \frac{\sigma\eta}{\Gamma(1-\sigma)} s^{-\sigma-1} e^{-s} \mathbb{I}_{(0,+\infty)}(s) ds \ ,$$

and $0 \leq \sigma \leq 1$, $\eta \geq 0$. Moreover, the sequences $(P_i)_i$ and $(X_i)_i$ are independent, $(X_i)_i$ being i.i.d. from G_0. This process includes several well-known stochastic processes, namely the Dirichlet process if $\sigma = 0$, the Normalised Inverse Gaussian process if $\sigma = 1/2$ (Lijoi *et al.* [15]), and it degenerates on G_0 if $\sigma = 1$. See also Pitman [18, 19].

Generally, the finite dimensional distributions of P are not available in closed analytic form. However, the distribution G_0 functions as the mean distribution of the process, because, for any set B,

$$\mathbb{E}\,(G(B)) = G_0(B) \quad \text{and} \quad \mathrm{Var}\,(G(B)) = G_0(B)(1 - G_0(B))\mathcal{I}(\sigma, \eta)$$

where $\mathcal{I}(\sigma, \eta)$ ranges in $(0, 1]$ and is a decreasing function of σ for fixed $\eta > 0$ and a decreasing function of η for fixed $\sigma \in (0, 1)$. See Argiento *et al.* [2] for an explicit expression of $\mathcal{I}(\sigma, \eta)$.

According to (4), the hyperparameters are G_0, η, σ and those within the priors of β_1.

We may draw an interesting analogy with the Bayesian mixed-effects model in Leon *et al.* [14], where the random effect parameters $(\alpha_1, \ldots, \alpha_8)$ in (2)-(3) are assumed exchangeable. In fact, conditionally on β_1, α_i and $\{(\vartheta_{1ij}, \vartheta_{2ij}), j = 1, \ldots, n_{xi}\}$, $i = 1, \ldots, 8$, model (4) is equivalent to

$$\begin{aligned} T_{xij} &\sim \mathrm{Weibull}(\vartheta_{1ij}, \lambda_{xij}) \quad j = 1, \ldots, n_{xi}, \ i = 1, \ldots, 8, \\ \log(\lambda_{xij}) &= \beta_1 \log(x) + \alpha_i + \log(\vartheta_{2ij}). \end{aligned}$$

The α_i's in the equation above are redundant, therefore we remove them, retaining only the vector $(\log \vartheta_{2ij})_{ij}$ of exchangeable parameters:

$$\begin{aligned} & T_{xij} \sim \mathrm{Weibull}(\vartheta_{1ij}, \lambda_{xij}) \quad j = 1, \ldots, n_{xi}, \ i = 1, \ldots, 8, \\ & \log(\lambda_{xij}) = \beta_1 \log(x) + \log(\vartheta_{2ij}) \\ & (\vartheta_{1ij}, \vartheta_{2ij}) \mid G \stackrel{iid}{\sim} G, \end{aligned} \tag{7}$$

with an NGG process prior for G, which takes the place of σ_α^2 in the Bayesian mixed-effects model. The terms $(\log \vartheta_{2ij})$ in (7) hold the same place of the α_i's in León *et al.* [14], with the difference that here the number of groups is random and can vary between one and $n = 108$, because of the ties induced by the discreteness of G. In this way we seem to have a twofold advantage over the Bayesian parametric mixed-effects model: the distribution of the error term is more flexible due to the non-parametric structure and we need not fix the number of groups and the membership to each group in advance, because they are inferred along with the other unknown quantities, thanks to the discreteness of G. With regards to this second point, in our specific problem two spools might have the same effect on survival for example, but we would not know in advance. We should observe that the components in the semiparametric mixture can be interpreted as latent groups in the data, but they are only loosely related to the levels taken by the spool effect. In fact, two lifetimes from the same spool could be assigned to different groups, maybe reflecting the influence of other unobserved latent factors. An alternative formulation of the model to handle the spool effects explicitly could be an AFT regression with parametric random effects for the spool and the error modeled as a non-parametric shape mixture of Weibull distributions. In any case the prior information concerning the actual number of spools can be incorporated into the model through (σ, η), since these hyperparameters induce a prior distribution on the number of clusters in (5). A sensitivity analysis can then follow by varying (σ, η) on a finite grid.

As a final remark on modeling issues, observe that, unlike in (3) where $\mathbb{E}(\alpha_i|\sigma_\alpha^2) = 0$, we have not assumed a random intercept β_0 in (7). If we did, we could rewrite $\log(\lambda_{xij})$ so that "random effect" has a zero conditional mean:

$$\log(\lambda_{xij}) = \beta_0 + \mathbb{E}(\log \vartheta_{2ij} \mid G) + \beta_1 \log(x) + \log(\vartheta_{2ij}) - \mathbb{E}(\log \vartheta_{2ij} \mid G);$$

in this case β_0 and $\mathbb{E}(\log \vartheta_{2ij} \mid G)$ would be confounded.

4 Data analysis

In this section we analyse the Kevlar fibre failure data and compare our results to those obtained previously by León *et al.* [14].

Model (4) needs to be completed with the specification of G_0 and the prior distribution for β_1. We assume β_1 to be normal with mean zero and variance 10^4. Concerning G_0 we choose the product of two independent gamma distributions, defining the prior density of $(\vartheta_1, \vartheta_2)$ as

$$g_0(\vartheta_1, \vartheta_2) = \frac{d^c}{\Gamma(c)} \vartheta_1^{c-1} \mathrm{e}^{-d\vartheta_1} \times \frac{b^a}{\Gamma(a)} \vartheta_2^{a-1} \mathrm{e}^{-b\vartheta_2}, \quad \vartheta_1 > 0,\ \vartheta_2 > 0.$$

Then we choose the hyperparameters a, b, c, d such that $\mathbb{E}(\log(V_{ij}))$ yields any desired prior mean for the intercept. As commonly done, we impose $\mathbb{E}(\log(V_{ij})) = 0$ for each i and j, provided hyperparameters are such that it exists. Straightforward conditioning arguments and the tables of integrals show that

$$\mathbb{E}(\log(V_{ij})) = \psi(a) - \log(b) - \gamma \frac{d}{c-1} \quad \text{if } c > 1,$$

and

$$\mathrm{Var}(\log(V_{ij})) = \frac{d^2}{(c-1)^2} \left[\left(\frac{\pi^2}{6} + \gamma \right) \frac{1}{c-2} + \frac{\pi^2}{6} \right] + \psi'(a) \quad \text{if } c > 2,$$

where $\gamma \simeq 0.577$ is the Euler constant and $\psi(\cdot)$ is the digamma function. We fixed several quadruplets of hyperparameters, with $c/d = 1$, representing the prior expected value of the shape parameter ϑ_2, whereas (a, b) is selected to have $\mathbb{E}(\log(V_{ij})) = 0$ when it is finite. Here we display results for $a = 0.5$, $b = 0.044$, $c = d = 2$ so that only the first moment of $\log V_{ij}$ is finite. In Table 1 we report point and interval estimates of the quantiles of the predictive distributions under our mixture model and compare them to those in the parametric Bayesian mixed-effects model in León *et al.* [14] (i.e. the median at the extrapolated stress level of 22.5 MPa, and the 0.01 quantile at the lowest stress level of 23.4 MPa of the predictive distributions). The distributions of the quantiles themselves are estimated from the sequence of quantiles of the G distribution sampled through the MCMC algorithm described in the Appendix. Notice how much narrower the interval estimates are compared to those under the parametric mixed-effects model, and how the predictive median survival times at stress level equal to 22.5 are much larger than those under the parametric model. We conjecture that this happens because the residual term in the semiparametric model has heavier tails, as implied, for instance, by the prior assumption that only the first moment of $\log V_{ij}$ is finite. In this case, the resulting prior for the error term yields a robust heavy-tailed semiparametric regression model. The values of hyperparameters (σ, η) governing the non-parametric mixing measure G were fixed so that the prior expected number of components in the mixture (also called *clusters*) is close to 8, in order to use the prior information on the number of spools.

Table 1. Median of the predictive distribution of failure time at stress 22.5 MPa and 0.01 quantile of the predictive distribution of failure time at stress 23.4 MPa (with 95% credibility intervals) under the semiparametric mixture model for two sets of hyperparameters, and under the parametric mixed-effects model

	$\sigma = 0.1, \eta = 10$	$\sigma = 0.3, \eta = 1$	*parametric mixed-effects*
Median 22.5[a]	204.84 (98.34, 406.86)	198.73 (93.01, 414.27)	53.68 (1.87, 1479)
0.01-quantile 23.4[b]	699.85(309.47,1776.50)	666.85(295.69,1700.17)	671 (21.96,19290)

[a] In thousands of hours; [b] in hours.

The plots of the posterior distributions of the number of clusters for (σ, η) equal to (0.1, 10) or (0.3, 1) (see Figure 1) have their modes in 5, 6 or 7, suggesting a more parsimonious modelling, with 8 still being an a posteriori credible value.

In Figure 2 a scatterplot of the log survival-time against the covariate (log-stress) is shown together with the estimates of the median survival time under the parametric and semiparametric AFT model. The regression line obtained under the frequentist AFT regression with Weibull errors agrees with our interval estimate. We notice that the semiparametric structure of the distribution of $\log(V)$ better follows the log-linear relationship between survival time and stress.

Finally, we consider other goodness-of-fit tools. The first are the *posterior predictive p-values*, as meant in Gelman *et al.* [7] Chapter 6, for the parametric and semiparametric mixed-effects models; more specifically, for all non-censored obser-

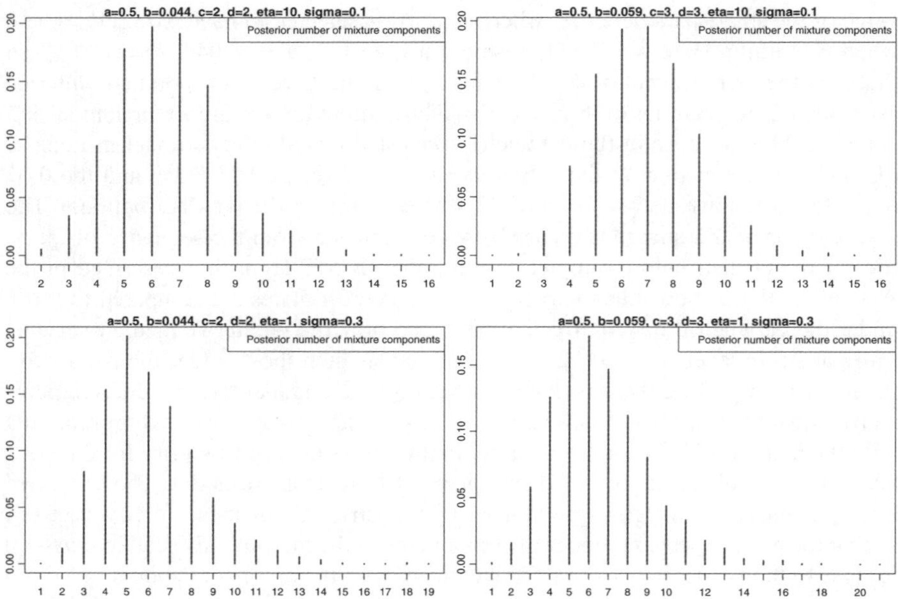

Fig. 1. Posterior distributions of the number of clusters under the NGG-mixture for different hyperparameter values

vations, we computed

$$\min\left(\mathbb{P}(T_i^{new} > t_i | data, x_i, spool_i), \mathbb{P}(T_i^{new} < t_i | data, x_i, spool_i)\right), \qquad i = 1, \ldots, 97$$

under the parametric model, using the WinBUGS code available in Leon's paper [14], averaging the tail probabilities given by (2), with respect to the MCMC iterate values of the parameters $(\beta_0, \beta_1, r, \alpha_i)$ in (3). Here T_i^{new} denotes the i-th "replicated data that could have been observed, or, to think predictively, as the data that we would see tomorrow if the experiment that produced t_i today were replicated with the same model and the same value of θ that produced the observed data" (Gelman *et al.* [7], Section 6.3). A value close to 0 denotes that the model is inadequate in fitting this observation. To simplify, observations with posterior predictive p-values less than 0.1 (let's say) were classified as "unusual" or "outlier".

Recall that our model (4) implicitly includes a random effect whose distribution is induced by the non-parametric prior. Therefore we calculate, for each i corresponding to an uncensored observation, the posterior predictive probability that the i-th replicated data T_i^{new} is larger than t_i

$$\int_{\mathbb{R}\times\Theta} \exp\left(-\left(\frac{t_i}{\lambda_{xi}}\right)^{\vartheta_{1i}}\right) \mathcal{L}(d\beta_1, d\theta_i | data),$$

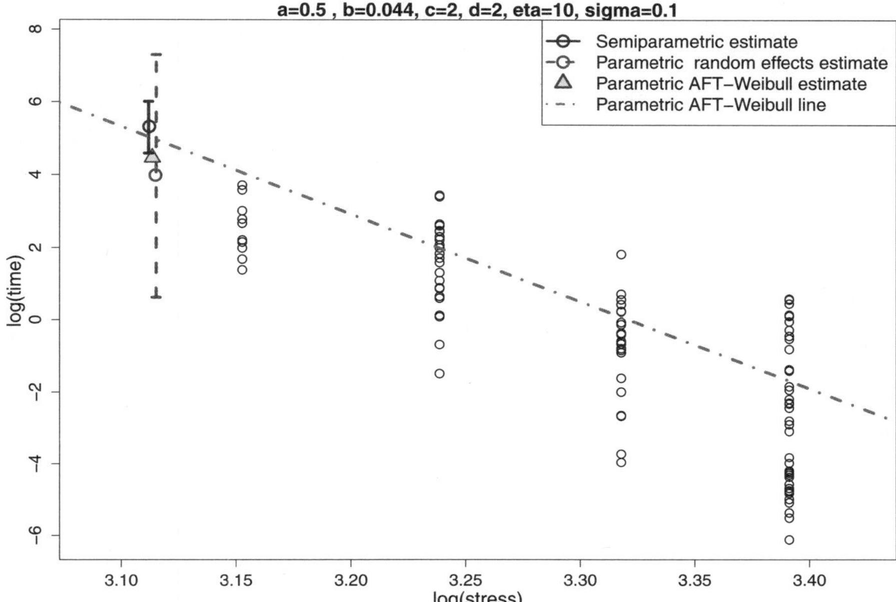

Fig. 2. Scatterplot of log-survival time against log-stress. Corresponding to log-stress equal to $\log(22.5) = 3.11$ the credible intervals of the predictive survival median time are shown (*dashed* is parametric, *solid* is semiparametric). The *dot-dashed* line is the parametric AFT regression line under Weibull errors, and the *triangle* is the corresponding estimate of the median

or smaller than t_i, and then we take the minimum. Here $\mathcal{L}(d\beta_1, d\theta_i|data)$ denotes the joint posterior of β_1 and the latent θ_i associated with t_i (check notation at the beginning of the Appendix). In Figure 3 we plotted the posterior predicted p-values just described, together with straight lines at level 0.1. Both models provide a good fit; the "unusual" observations under the parametric mixed-effects model are 18, but they are only 4 in our setting. As for the interval estimates, we explain the good performance of our p-values by the heavy-tailed error distributions.

The second goodness-of-fit tool we computed are the Bayesian residuals, as suggested in Chaloner and Brant [4], under the two models. In that paper, the authors define an outlier as any observation which has a surprising value of the "error" W_{ij}. Equation (1), which is the likelihood for the parametric model, can be written as

$$\log T_{xij} = \beta_0 + \beta_1 \log x + \alpha_i + \frac{W_{ij}}{r}, \; j = 1, \ldots, n_{xi}, \; i = 1, \ldots, 8, \qquad (8)$$

where, conditioning on the parameters and the covariates, the W_{ij}'s are independent and identically distributed according to the (standard) Gumbel distribution function

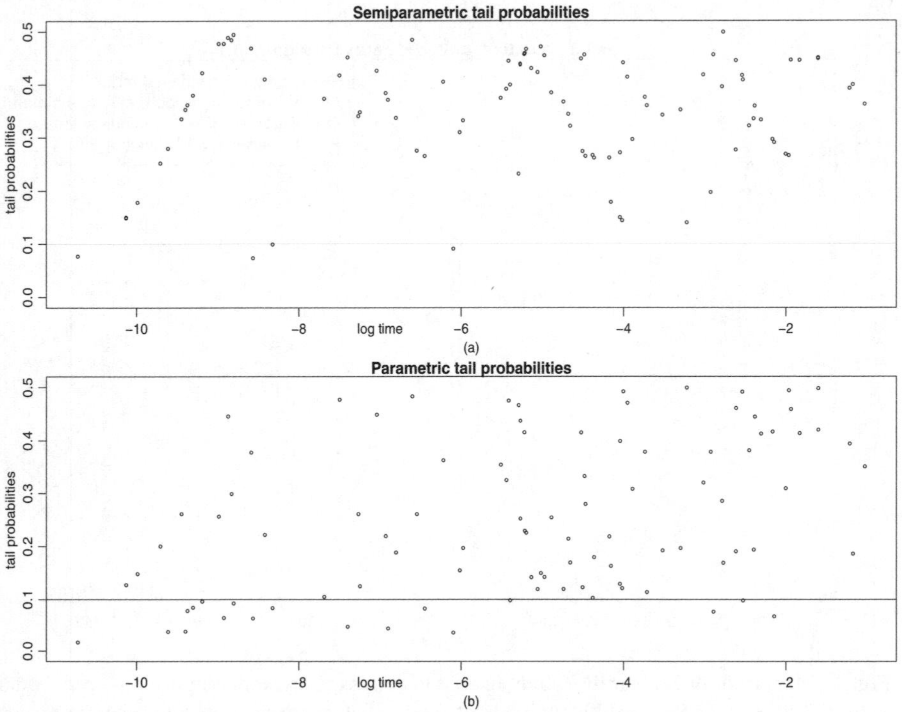

Fig. 3. Predictive tail probabilities for the uncensored observations under (**b**) parametric and (**a**) semiparametric models; straight lines are at level 0.1

$F_W(w) = 1 - \exp(-\exp(w))$. Therefore we consider the *realised* errors (for uncensored observations only)

$$\varepsilon_{ij} = r\left(\log t_{ij} - (\beta_0 + \beta_1 \log x_{ij} + \alpha_i)\right),\ j = 1, \ldots, n_{xi},\ i = 1, \ldots, 8,$$

obtained solving (8) w.r.t. W_{ij}. On the other hand, under the non-parametric model, (7) is equivalent to

$$\log T_{xij} = \beta_1 \log x + \log(\vartheta_{2ij}) + \frac{W_{ij}}{\vartheta_{1ij}},\ j = 1, \ldots, n_{xi},\ i = 1, \ldots, 8,$$

so that the Bayesian residuals for uncensored observations are

$$\epsilon_{ij} = \vartheta_{1ij}\left(\log t_{ij} - (\beta_1 x_i + \log \vartheta_{2ij})\right).$$

Each ε_{ij} is a function of the unknown parameters, so that its posterior distribution can be computed through MCMC simulation, and later examined for indications of possible departures from the assumed model and the presence of outliers (see also Chaloner [3]). Therefore, it is sensible to plot credibility intervals for the marginal

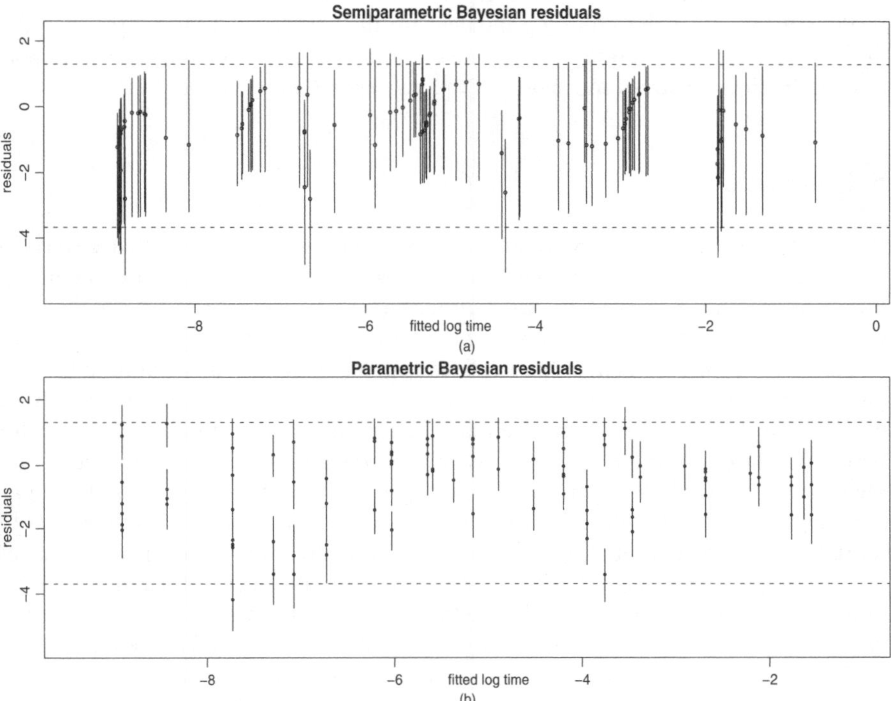

Fig. 4. Bayesian residuals for the uncensored observations under (**b**) parametric and (**a**) semiparametric models; the *dashed* lines are 0.025- and 0.975-quantiles of the standard Gumbel distribution

posterior of each ε_{ij}, comparing them to the marginal prior credibility intervals (of the same level).

For the plotting we computed the fitted log-failure times under the two models:

$$\widehat{\log T_{xij}} = \mathbb{E}\left(\beta_1 \log x_i + \alpha_i + \frac{W_{ij}}{r}|data\right) \quad \text{(parametric)}$$

$$\widehat{\log T_{xij}} = \mathbb{E}\left(\beta_1 \log x_i + \log(\vartheta_{2ij}) + \frac{W_{ij}}{\vartheta_{1ij}}|data\right) \quad \text{(semiparametric).}$$

Figure 4 displays the 95% credibility intervals of the Bayesian residuals under the semiparametric and parametric models respectively, against the "fitted" log-failure times. The dashed lines in the plot denote the 0.025 and 0.975 quantiles of the prior marginal distribution (the standard Gumbel). The medians of the residuals are less close to the prior limits in the semiparametric model; on the other hand the credibility intervals are wider, because of the additional variability determined by the random partitioning of the observations. The residuals are also less scattered, which is probably due to a re-classification of observations into more homogeneous groups (with respect to their mean) than the grouping induced by the spools. In Figure 4(b), we also observe that the distinct fitted log-failure times are 27, corresponding to the number of combinations of stress and spool for the uncensored observations.

5 Conclusions

We have fitted a dataset of 108 lifetimes of Kevlar pressure vessels with two covariates (spool and stress) to an AFT model. The error term was modeled as a non-parametric hierarchical mixture of Weibull distributions on both the shape and the rate, yielding an alternative to a parametric Bayesian mixed-effects model. Here the mixing measure was assumed to be a normalised generalised gamma random measure, indexed by two parameters. The goodness-of-fit of the parametric and the semiparametric models are comparable. We also found that the latter better follows the log-linear relationship between failure times and covariates. We obtained narrower interval estimates of the quantiles considered. As mentioned in the paper, our model presents an analogy with the Bayesian parametric mixed-effects model; in any case, its advantages consist of a more flexible distribution of the error term and a random number of groups induced by the non-parametric model itself. However, one should be aware that the grouping of the observations embeds both the random spool effect and possibly of other latent effects. In our opinion, this is not necessarily a drawback, because the information on the spool effect is included in the hyperparameters of the non-parametric process prior. Hence, as long as the model shows good predictive performances, both known and unknown random effects are duly taken into account.

Acknowledgement. We thank Annalisa Cadonna for running the WinBUGS code and for computing the Bayesian estimates of the parametric mixed-effects model of León *et al.* [14].

Appendix

Computation of full Bayesian inferences requires knowledge of the posterior distribution of the random distribution function $F(v; G)$ in (5), as well as the posterior distribution of the regression parameter β_1. Here it is possible to build an MCMC algorithm which approximates the posterior distribution of G, so that we will also provide credibility intervals for $F(v; G)$.

The hierarchical structure of model (4), indicates that, conditionally on G, every lifetime T_i is associated with a point $\theta_i = (\vartheta_{1i}, \vartheta_{2i})$ from the support of G. These points are not necessarily distinct. We denote the distinct values within the n-tuple $\underline{\theta} = (\theta_1, \ldots, \theta_n)$ by $\underline{\psi} = (\psi_1, \ldots, \psi_{n(\pi)})$, where $n(\pi)$ is their number, $1 \leq n(\pi) \leq n$. The elements of $\underline{\psi}$ are matched to the elements of $\underline{\theta}$ by means of the induced partition of the indexes $\{1, \ldots, n\}$, which we denote by $\pi = \{C_1, \ldots, C_{n(\pi)}\}$, where $C_j = \{i : \theta_i = \psi_j\}$.

Let $t_{98}^{108} = (t_{98}, \dots, t_{108})$ be the vector of the imputed censored failure times. The state of the Markov chain will be $(G, \underline{\theta}, \beta_1, t_{98}^{108}) = (G, \underline{\psi}, \pi, \beta_1, t_{98}^{108})$. In order to build a Gibbs sampler, we must be able to sample from all the full conditional posterior distributions, and in particular from the full conditional of G. By a characterisation of the posterior distribution of G given in James *et al.* [9], sampling from the full conditional of G amounts to sampling the $n(\pi)$ weights assigned to the points in $\underline{\psi}$ and both the (infinite) remaining weights and support points of G. An augmentation of the state space with an auxiliary variable u, allows for an independent conditional sampling of the two groups of random variables. Then the actual state of the chain will be $(G, \underline{\psi}, \pi, \beta_1, t_{98}^{108}, u)$. For more explanation and details concerning the meaning of u, we refer the reader to James *et al.* [9], Nieto-Barajas and Prünster [17] and to Argiento *et al.* [2] and here simply sketch the steps of our algorithm, with the square bracket notation denoting probability distributions. We also omit the indication of the observed failure times among the conditioning random variables for ease of notation.

Sampling G: By conditional independence, G depends on the observed lifetimes only through the vector $\underline{\theta}$. Therefore, using the equivalent representation $(\underline{\psi}, \pi)$ for $\underline{\theta}$, $[G \mid \underline{\psi}, \pi, u]$ is the law of the following random measure (to be normalised)

$$G^* \propto \sum_{j=1}^{+\infty} J_j \delta_{\tau_j} + \sum_{j=1}^{n(\pi)} L_j \delta_{\psi_j},$$

where the ψ_j's are fixed points in the support of G^* and the remaining weights and support points are random. While the sequence $(J_j)_j$ should be infinite, we use a finite sequence $(J_j)_{1 \le j \le M}$, where the truncation point M is chosen as described in Argiento *et al.* [2], along with the details of the simulation method from the Poisson process.

Sampling $\underline{\theta}$: Using the truncated distribution G^* sampled at the previous step, we have, as $i = 1, \dots, n$, $[\theta_i \mid G^*, \underline{\theta}_{-i}, \beta_1, t_{98}^{108}, u] = [\theta_i \mid G^*, \beta_1, t_{98}^{108}]$, where the dependence on the failure times and β_1 is retained, because θ_i is the parameter of the kernel density. Using Bayes theorem it can be shown that, for $i = 1, \dots, n$,

$$[\theta_i \mid G^*, \beta_1, t_{98}^{108}] \propto \sum_{j=1}^{M} J_j k(v_i; \theta_i) \delta_{\tau_j}(d\theta_i) + \sum_{j=1}^{n(\pi)} L_j k(v_i; \theta_i) \delta_{\psi_j}(d\theta_i),$$

where $v_i = \mathrm{e}^{-\beta_1 \log(x_i)} t_i$.

Sampling β_1: The full conditional of β_1 is log-concave and is amenable to adaptive rejection sampling.

Sampling u: James *et al.* [9] derive the full conditional of u, which is a non-standard distribution, and a Metropolis step was expressly designed.

Sampling t_{98}^{108}: All the censored failure times have the same censoring point at 41,000 hours. The failure times are conditionally independent given $\underline{\theta}$, so that the full conditional of t_i, $i = 98, \dots, 108$ is a left-truncated Weibull distribution with shape ϑ_{1i} and rate ϑ_{2i}, so that its survival function can be inverted exactly. This concludes the MCMC sweep.

References

1. Argiento, R., Guglielmi, A., Pievatolo, A.: A comparison of nonparametric priors in hierarchical mixture modelling for AFT regression. J. Statist. Plann. Inference **139**, 3989–4005 (2009)
2. Argiento, R., Guglielmi, A., Pievatolo, A.: Bayesian density estimation and model selection using nonparametric hierarchical mixtures. Comput. Statist. Data Anal. **54**, 816–832 (2010)
3. Chaloner, K.: Bayesian residual analysis in the presence of censoring. Biometrika **78**, 637–644 (1991)
4. Chaloner, K., Brant, R.: A Bayesian approach to outlier detection and residual analysis. Biometrika **31**, 651–659 (1988)
5. Crowder, M.J., Kimber, A.C., Smith, R.L., Sweeting, T.J.: Statistical Analysis of Reliability Data. Chapman & Hall/CRC, Boca Raton, Florida (1991)
6. Gelfand, A.E., Kottas, A.: Bayesian semiparametric regression for median residual life. Scand. J. Statist. **30**, 651–665 (2003)
7. Gelman, A., Carlin, J.B., Stern, H.S., Rubin, D.B.: Bayesian Data Analysis. second edn., Chapman & Hall/CRC, Boca Raton, Florida (2004)
8. Hanson, T.E.: Modeling censored lifetime data using a mixture of gammas baseline. Bayesian Anal. **1**, 575–594 (2006)
9. James, L.F., Lijoi, A., Pruenster, I.: Posterior Analysis for Normalized Random Measures with Independent Increments. Scand. J. Statist. **36**, 76–97 (2008)
10. Jara, A., Lesaffre, E., de Iorio, M., Quintana, F.A.: Bayesian Semiparametric Inference for Multivariate Doubly-Interval-Censored Data. Technical report (2008)
11. Kleinman, K.P., Ibrahim, J.G.: A semiparametric Bayesian approach to generalized linear mixed models. Stat. Med. **17**, 2579–2596 (1998)
12. Kottas, A., Gelfand, A.E.: Bayesian semiparametric median regression modeling. J. Amer. Statist. Assoc. **96** 1458–1468 (2001)
13. Kuo, L., Mallick, B.: Bayesian semiparametric inference for the accelerated failure-time model. Can. J. Statist. **25**, 475–472 (1997)
14. León, R.V., Ramachandran, R., Ashby, A.J., Thyagarajan, J.: Bayesian Modeling of Accelerated Life Tests with Random Effects. J. Qual. Technol. **39**, 3–13 (2007)
15. Lijoi, A., Mena, R.H., Prünster, I.: Hierarchical mixture modeling with normalized inverse-Gaussian priors. J. Amer. Statist. Assoc. **100**, 1278–1291 (2005)
16. Lo, A.: On a class of Bayesian nonparametric estimates I. Density estimates. Ann. Statist. **12**, 351–357 (1984)
17. Nieto-Barajas, L.E., Prünster, I.: A sensitivity analysis for Bayesian nonparametric density estimators. Statist. Sinica **19**, 685–705 (2009)
18. Pitman, J.: Some Developments of the Blackwell-MacQueen Urn Scheme. In: Ferguson T.S. et al. (eds.) *Statistics, Probability and Game Theory; Papers in honor of David Blackwell, volume 30 of Lecture Notes-Monograph Series*. Institute of Mathematical Statistics, Hayward, California, 245–267 (1996)
19. Pitman, J.: Poisson-Kingman partitions. In: Goldstein, D.R. (eds.) *Science and Statistics: a Festschrift for Terry Speed*. Institute of Mathematical Statistics, Hayward, California, 1–34 (2003)

Space filling and locally optimal designs for Gaussian Universal Kriging

Alessandro Baldi Antognini and Maroussa Zagoraiou

Abstract. Computer simulations are often used to replace physical experiments aimed at exploring the complex relationships between input and output variables. Undoubtedly, computer experiments have several advantages over real ones, however, when the response function is complex, simulation runs may be very expensive and/or time-consuming, and a possible solution consists of approximating the simulator by a suitable stochastic metamodel, simpler and much faster to run. Several metamodel techniques have been suggested in the literature and one of the most popular is the Kriging methodology. In this paper we study the optimal design problem for the Universal Kriging metamodel with respect to different approaches, related to prediction, information gain and estimation. Also we give further justifications and some criticism concerning the adoption of the space filling designs, based on theoretical results and numerical evidence as well.

Key words: simulator, metamodel, Universal Kriging, D-optimality, space filling design, entropy, integrated mean squared prediction error

1 Introduction

Processes are so complex in many areas of science and technology that physical experimentation is often too time-consuming or expensive. Thus, computer simulations are often used to replace real experiments. These experiments rely on several runs of a computer code, which implements a simulation model of a physical system of interest, usually aimed at investigating a deterministic Input/Output (I/O) relationship. However, the request that the simulator should be accurate in describing the physical system means that the simulator itself may be rather complex. This has led to the use of surrogate models, also called metamodels, which are computationally cheaper alternatives of the simulation models. In general, these metamodels are statistical interpolators, fitted from the simulated data, which represent a valid approximation of the original simulator and allow us to obtain efficient prediction of the I/O function at the entire design region. Several surrogate techniques have been suggested in the literature and the Kriging methodology, introduced by D.G. Krige and further developed by different authors, has become one of the most popular in the context of

Mantovan, P., Secchi, P. (Eds.): Complex Data Modeling and Computationally Intensive Statistical Methods

computer experiments since the pioneering paper of Sacks *et al.* [14]. Adopting this approach, the deterministic output of the computer code is regarded as a realisation of a stationary Gaussian process, whose covariance structure is usually modeled in a parametric way to reflect the available information about the I/O function. After data have been collected, the unknown parameters of the model will be estimated and the Kriging interpolator will be derived in order to predict the deterministic output at untried sites.

In this setting, the experimental design problem consists of selecting the inputs at which to run a computer code. *Space filling* and *optimal criterion-based* designs are the main approaches found in the literature for designing a computer experiment. Space-filling designs represent the most widely used class of procedures in this application field. The reason lies in their simplicity and the fact that it is often important to obtain information from the entire design space, which has to be filled in a uniform way. There are several techniques to define what it means to spread points, especially in the multidimensional setting, and these lead to a variety of designs within the space filling class, such as the well-known Latin hypercube designs (for recent literature see [9]).

Adopting an optimal design approach, given the metamodel it is possible to formulate specific optimality criteria which reflect the experimental goals (see [5]). For instance, the choice of the observation points could be carried out in order to acquire the maximum amount of information from the experiment. Assuming an information-theoretic approach, Shannon's entropy has been widely used in statistics as a measure of uncertainty and the change in entropy before and after collecting data was introduced as a design criterion with the aim of maximising the information gain (see for instance [16]). Otherwise, the input locations can be chosen for accurate prediction of the output of the simulator and most of the progress in this area has been made with respect to criteria based on suitable functionals of the *Mean Square Prediction Error* (MSPE), such as the *Integrated Mean Squared Prediction Error* representing an averaged version of the MSPE over the design space (see for example [15]). Recently, several authors have focused on the problem of choosing the design in order to estimate the parameters of interest with high efficiency. Since the Fisher information matrix can still be regarded as a measure of the estimation precision even in the presence of correlated observations, classical optimality criteria (like for instance D, D_s, ...) have been taken into account [10,22,24].

The aim of this paper is to analyse the optimal design problem for the Universal Kriging metamodel with respect to different approaches, related to prediction, information gain and estimation, generalising some recently obtained results by [4,7,12,17,18,22]. Starting from a brief discussion about the optimality properties of the space filling designs for the Ordinary Kriging, in Section 3 we prove that these procedures are still optimal for the Universal Kriging with respect to the entropy criterion. Section 4 deals with the optimal designs for estimation and prediction showing that, in general, these criteria lead to local optimality problems. However, even in this critical situation, numerical evidence points out that the adoption of the space filling design is still rationale except in the case of the D-optimality criterion, namely when the inferential aim is focused on the joint estimation of all the unknown parameters. Furthermore, the space filling designs perform particularly well when the correlation between the observations is medium/high and, at the same time, the loss of efficiency induced by these designs is quite small even when the observations tend to be uncorrelated, i.e. in the case of misspecification of the model. Finally, Section 5 is dedicated to some conclusions and further developments in this topic.

2 Kriging methodology

Following [14], the Kriging approach consists of assuming the output $y(x)$ of the simulator as a realisation of a Gaussian process $Y(x)$, also called Gaussian random field, in the form

$$Y(x) = \mu(x) + Z(x), \quad x \in \Xi \subset \mathbb{R},$$

where $\mu(x)$ denotes the trend component, Ξ is a compact interval of $\mathbb{R}$ and $Z(x)$ represents the departure of the response variable $Y(x)$ from the trend. More precisely, $Z(\cdot)$ is a zero-mean stationary Gaussian process with constant marginal variance σ^2 and non-negative correlation function between two inputs that depends on their distance and tends towards 1 as the distance moves towards 0.

Two different types of Kriging metamodels have been proposed in the literature depending on the functional form of the trend component:

- *Universal Kriging*: the trend depends on x and is modeled in a regressive way:

$$\mu(x) = f(x)^t \beta, \tag{1}$$

  where $f(x) = (f_1(x), \dots, f_p(x))^t$ is a known vector function and β is a p-dim vector of regressor coefficients;
- a special case of (1) is the widely used *Ordinary Kriging*, characterised by a constant but unknown trend:

$$\mu(x) = \beta_0 \quad \text{for any } x \in \Xi. \tag{2}$$

The study of Gaussian random fields is also a study of covariance or correlation functions since they should be chosen in such a way to reflect the characteristics of the output. For instance, for a smooth response, a correlation function with some derivatives would be preferred, while an irregular response might call for a function with no derivatives. A customary choice can be found within the power exponential family, which is the most common model in the computer experiment literature. In the 1-dimensional case, this class of correlation functions is given by

$$\text{corr}\{Z(x_i), Z(x_j)\} = e^{-\theta|x_i - x_j|^p} \quad x_i, x_j \in \Xi,$$

where θ is a non-negative parameter that governs the intensity of the correlations between input points and $p \in (0; 2]$ is a smoothing parameter: for $p = 2$ we obtain the Gaussian correlation, suitable for smooth and infinitely differentiable responses, while when $p = 1$ we have the so-called exponential correlation. For a thorough description of this topic see [1].

In this paper we will focus our attention on the Universal Kriging (3) in the form

$$Y(x) = \beta_0 + \beta_1 x + Z(x), \quad x \in \Xi \subset \mathbb{R}, \tag{3}$$

with exponential correlation structure

$$\text{corr}\{Z(x_i), Z(x_j)\} = e^{-\theta|x_i - x_j|} \quad \theta > 0, \quad x_i, x_j \in \Xi, \tag{4}$$

and we will consider the problem of an n-points design $\xi = \{x_1, x_2, \ldots, x_n\}$, in order to reduce in some sense the uncertainty of the above metamodel. In what follows, we set (w.l.o.g.) $\Xi = [0, 1]$ and $x_1 < x_2 < \ldots < x_n$ (the design points need to be distinct, due to the deterministic nature of the simulator). Furthermore, we also assume that $x_1 = 0$ and $x_n = 1$, since under Kriging methodology, predictions in the case of extrapolation are not advisable.

Observe that every n-points design induces univocally an $(n-1)-$dim vector of distances $(d_1, \ldots, d_{n-1})$ where

$$d_i = x_{i+1} - x_i \in (0; 1) \quad \text{for any } i = 1, \ldots, n-1 \quad \text{and} \quad \sum_{i=1}^{n-1} d_i = 1,$$

and viceversa. Thus, from now on, ξ will denote the design in terms of both the set of input locations $0 = x_1 < x_2 < \ldots < x_n = 1$ or the corresponding vector of distances $(d_1, \ldots, d_{n-1})$.

3 Optimality of space filling designs

Space filling designs are very popular in the computer experiment literature, since they represent a very natural and intuitive choice (for a thorough description, see [15]). When the design space Ξ is a compact subset of $\mathbb{R}$, a space filling design with n points $\bar{\xi}$, also called *equidistant*, is generated by a set of equispaced input locations, namely

$$x_i = \frac{i-1}{n-1}, \quad \text{for any} \quad i = 1, \ldots, n,$$

or in term of distances

$$d_1 = d_2 = \ldots = d_{n-1} = (n-1)^{-1} \quad \text{for any} \quad n \geq 3.$$

However, the vast use of this class of procedures lies not only in their simplicity; indeed, as shown by Baldi Antognini and Zagoraiou [4] for the Ordinary Kriging model, the adoption of the equispaced design can be often regarded as an optimal solution as the next theorem shows.

Theorem 1. *(Baldi Antognini and Zagoraiou, 2009) Under the Ordinary Kriging (2) with exponential correlation structure (4), for any sample size, the equidistant design* $\bar{\xi}$*:*

- *maximises the expected reduction in entropy;*
- *minimises the Integrated Mean Squared Prediction Error;*
- *maximises the information of the constant trend.*

Therefore, the space filling design is optimal with respect to the information gain, and also concerning the prediction and estimation of β_0. However, $\bar{\xi}$ minimises the information of the correlation parameter and thus it represents the worst possible solution for the estimation of θ. Moreover, when the inferential aim consists of estimating both parameters β_0 and θ, the D-optimal design depends itself on θ which leads to a locally optimal solution.

A straightforward consequence of the previous theorem is that the space filling design is still optimal under the Universal Kriging with respect to the information gain, as shown by the next corollary.

Corollary 1. *Under the Universal Kriging stated in (3) and (4), for any given sample size the space filling design $\bar{\xi}$ is optimal with respect to the entropy criterion.*

Proof. Since the entropy is not affected by the structure of the trend component, the result follows directly from the proof of theorem 5.2 in [4].

4 Locally optimal designs for Universal Kriging

4.1 Optimal designs for estimation

The problem arises of an optimal choice of the design points in order to estimate the unknown parameters of the model with maximum precision. Assuming the maximum likelihood approach, as shown by several authors (see for instance [2, 11, 23]) the inverse of the Fisher information is a reasonable approximation of the covariance matrix of the MLE's, even in the presence of correlated observations; so, it is quite natural to extend the classic optimal design theory to the present setting by taking into account the classical optimality criteria, namely suitable functional of the Fisher information matrix aimed at combining the uncertainty of the parameters of interest (see for example [3]). Of special interest in our context are the D- and D_s-optimality criteria adopting in this section.

Recalling that in the case of 1-dimensional input space the covariance parameters θ and σ^2 cannot be identified simultaneously [21], in what follows, the variance will be considered as a nuisance parameter and for simplicity we let (wlog) $\sigma^2 = 1$, whereas $\beta = (\beta_0; \beta_1)^t$ and θ are the parameters of interest.

Let $Y_i = Y(x_i)$ and $f(x_i) = (1, x_i)^t$ for any $i = 1, \ldots, n$, then the (3×3) Fisher information associated with a design ξ is given by:

$$I(\beta, \theta; \xi) = \begin{pmatrix} F^t C^{-1} F & 0 \\ 0^t & \frac{1}{2}\operatorname{tr}\left[\left(C^{-1}\frac{\partial C}{\partial \theta}\right)^2\right] \end{pmatrix},$$

(see for instance [11]), where $F = (f(x_1), \ldots, f(x_n))^t$ and $C = C(\theta)$ represents the correlation matrix of the observations $Y_1, \ldots, Y_n$; from (4), C is an $(n \times n)$ symmetric matrix whose (i, j)-th element is $e^{-\theta|x_i - x_j|}$, which depends on the corresponding design points only through their distance.

Firstly, we give a simplification of the Fisher information matrix.

Theorem 2. *Under the Universal Kriging model stated in (3) and (4), the Fisher information $I(\beta, \theta; \xi)$ can be written as follows*

$$I(\beta, \theta; \xi) = \begin{pmatrix} I(\beta;\xi) & 0 \\ 0^t & I(\theta;\xi) \end{pmatrix}, \tag{5}$$

where

$$I(\beta;\xi) = \begin{pmatrix} \sum_{i=1}^{n-1}\left(x_{i+1} - x_i + \frac{1-e^{-\theta(x_{i+1}-x_i)}}{1+e^{-\theta(x_{i+1}-x_i)}}\right) & \sum_{i=1}^{n-1} \frac{x_{i+1}-x_i e^{-\theta(x_{i+1}-x_i)}}{1+e^{-\theta(x_{i+1}-x_i)}} \\ \sum_{i=1}^{n-1} \frac{x_{i+1}-x_i e^{-\theta(x_{i+1}-x_i)}}{1+e^{-\theta(x_{i+1}-x_i)}} & \sum_{i=1}^{n-1} \frac{\left(x_{i+1}-x_i e^{-\theta(x_{i+1}-x_i)}\right)^2}{1-e^{-2\theta(x_{i+1}-x_i)}} \end{pmatrix}, \tag{6}$$

and

$$I(\theta;\xi) = \sum_{i=1}^{n-1} \frac{(x_{i+1}-x_i)^2\left(1+e^{2\theta(x_{i+1}-x_i)}\right)}{\left(e^{2\theta(x_{i+1}-x_i)}-1\right)^2}. \tag{7}$$

Proof. Following [22], the correlation matrix can be written as $C = LDL^t$, where

$$L = \begin{pmatrix} 1 & 0 & 0 & \ldots & 0 \\ e^{-\theta d_1} & 1 & 0 & \ldots & 0 \\ e^{-\theta \sum_{i=1}^{2} d_i} & e^{-\theta d_2} & 1 & \ldots & 0 \\ \vdots & \vdots & \vdots & \ddots & \vdots \\ e^{-\theta \sum_{i=1}^{n-1} d_i} & e^{-\theta \sum_{i=2}^{n-1} d_i} & e^{-\theta \sum_{i=3}^{n-1} d_i} & \ldots & 1 \end{pmatrix},$$

and D is an $(n \times n)$-diagonal matrix with $D = diag\left(1, 1-e^{-2\theta d_1}, \ldots, 1-e^{-2\theta d_{n-1}}\right)$. Then, $C^{-1} = (D^{-\frac{1}{2}}L^{-1})^t(D^{-\frac{1}{2}}L^{-1})$ where

$$L^{-1} = \begin{pmatrix} 1 & 0 & 0 & \ldots & 0 \\ -e^{-\theta d_1} & 1 & 0 & \ldots & 0 \\ 0 & -e^{-\theta d_2} & 1 & \ldots & 0 \\ \vdots & \ddots & \ddots & \ddots & \vdots \\ 0 & \ldots & 0 & -e^{-\theta d_{n-1}} & 1 \end{pmatrix},$$

and thus, $I(\beta;\xi) = F^t C^{-1} F = (D^{-\frac{1}{2}} L^{-1} F)^t (D^{-\frac{1}{2}} L^{-1} F)$ with

$$D^{-\frac{1}{2}} L^{-1} F = \begin{pmatrix} 1 & 0 \\ \frac{1-e^{-\theta d_1}}{\sqrt{1-e^{-2\theta d_1}}} & \frac{x_2 - x_1 e^{-\theta d_1}}{\sqrt{1-e^{-2\theta d_1}}} \\ \vdots & \vdots \\ \frac{1-e^{-\theta d_{n-1}}}{\sqrt{1-e^{-2\theta n-1}}} & \frac{x_n - x_{n-1} e^{-\theta d_{n-1}}}{\sqrt{1-e^{-2\theta d_{n-1}}}} \end{pmatrix}.$$

Analogously, the expression $I(\theta;\xi)$ can be derived by tedious calculations (see [22]).

Notice that, from (5), (6) and (7):

- both components $I(\beta;\xi)$ and $I(\theta;\xi)$ depend on θ;
- the Fisher information has a block-form, so that the estimation of the trend $\beta = (\beta_0, \beta_1)^t$ is uncorrelated with respect to that of the covariance parameters θ, while the estimations of β_0 and β_1 are correlated;
- contrary to the Ordinary Kriging (see [4]), the Fisher information matrix cannot be decomposed in an additive way in terms of information associated with each distance between adjacent points. In fact,

$$I(\beta,\theta;\xi) = \sum_{i=1}^{n-1} \begin{pmatrix} d_i + \frac{1-e^{-\theta d_i}}{1+e^{-\theta d_i}} & \frac{x_{i+1} - x_i e^{-\theta d_i}}{1+e^{-\theta d_i}} & 0 \\ \frac{x_{i+1} - x_i e^{-\theta d_i}}{1+e^{-\theta d_i}} & \frac{(x_{i+1} - x_i e^{-\theta d_i})^2}{1-e^{-2\theta d_i}} & 0 \\ 0 & 0 & \frac{d_i^2 (1+e^{2\theta d_i})}{(e^{2\theta d_i}-1)^2} \end{pmatrix}.$$

  Thus, recalling that $x_{i+1} = \sum_{j=1}^{i} d_j$ for any $i = 1, \ldots, n-1$, each design point makes a contribution to the Fisher information which depends on the distances between the point taken into consideration and all the previous input locations.

Since for the Universal Kriging the estimation of the trend plays a fundamental role, now we focus on the optimal design problem for estimating $\beta = (\beta_0, \beta_1)^t$. This corresponds to the adoption of the so-called D_s-optimality criterion (see for instance [3]) and the aim is to find the design that maximises

$$\frac{\det I(\beta,\theta;\xi)}{I(\theta;\xi)} = \det I(\beta;\xi). \tag{8}$$

For $n = 3$, i.e. in the case of only one design point $\xi = \{x_1 = 0, x_2, x_3 = 1\}$, numerical optimisation shows that the design that maximises $\det I(\beta;\xi)$ depends on the value of θ (local optimality), and from here will be indicated by $\xi_\theta^{D_s}$. In particular, in the presence of medium/high correlated observations (i.e. for small values of θ) the D_s-optimal design coincides with the equispaced one; while, as θ increases, $\xi_\theta^{D_s}$ moves towards the extremes of the interval (see also Figure 1).

Indeed, when the observations tend to be independent,

$$\lim_{\theta\to\infty} \det I(\beta;\xi) = 2(x_2^2 - x_2 + 1),$$

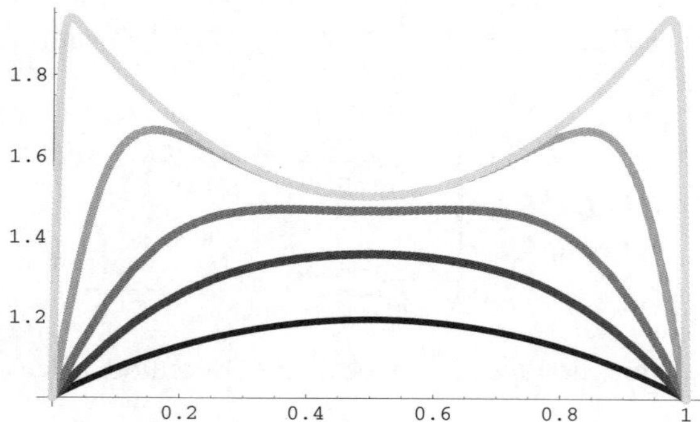

Fig. 1. D_s-optimality criterion for the estimation of the trend: x_2 varies on the x-axis. The values of θ from the bottom curve to the top curve are: 3, 5, 8, 20, 200, respectively

so that, according to the optimal design theory for the classical linear model, the optimal choice consists of putting the input at an extreme of the experimental region.

As the sample size increases, numerical maximisation reflects the same behaviour as previously, leading to a locally D_s-optimal design: the equispaced design is still optimal for small values of the correlation parameter, while as θ grows the supremum of $\det I(\beta; \xi)$ moves towards the frontier points of Ξ. For instance, when $n = 4$ we obtain

$$\lim_{\theta \to \infty} \det I(\beta; \xi) = 3(x_2^2 + x_3^2) - 2x_2(1 + x_3) - 2x_3 + 3.$$

Taking into account the efficiency, it is possible to compare the performances of the space filling design $\bar{\xi}$ with respect to the locally optimal one $\xi_\theta^{D_s}$ for different values of θ. From (8) and (6), the D_s-efficiency is given by

$$D_s\text{-Eff} = \left(\frac{\det I(\beta; \bar{\xi})}{\det I\left(\beta; \xi_\theta^{D_s}\right)} \right)^{\frac{1}{2}}.$$

As can be seen in Figure 2, the space filling design performs very well even in the case of local optimality for every value of the correlation parameter, since it is D_s-optimal for small θ's, and, at the same time, guarantees high efficiency as θ increases.

Regarding the estimation of the correlation parameter, the D_s-optimality criterion for the Universal Kriging coincides with that of the Ordinary Kriging, since $I(\theta; \xi)$ is not affected by the trend structure. Thus, as shown in [22], the optimal design does not exist and the approximated D_s-optimal procedure for the estimation of θ tends to put the observations at the extremes of the interval.

When the experimental aim consists of minimising the uncertainty regarding the joint estimation of the whole set of parameters of interest β_0, β_1 and θ, it is customary to assume the D-optimality criterion, so that the problem is to find the design that maximises

$$\det I(\beta, \theta; \xi) = I(\theta; \xi) \cdot [\det I(\beta; \xi)].$$

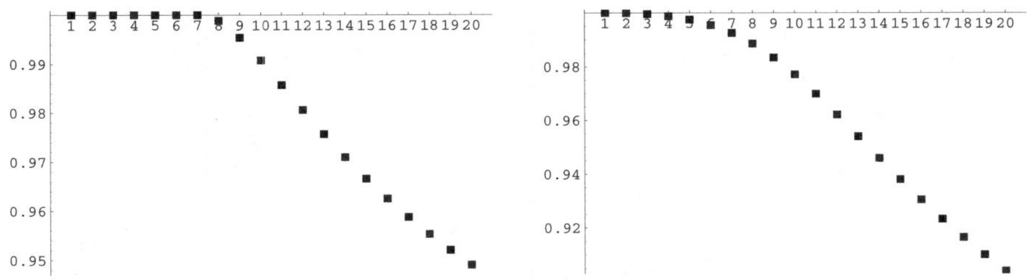

Fig. 2. Plots of the D_s-efficiency for the trend parameter $\beta = (\beta_0, \beta_1)^t$ as θ varies with $n = 3$ (left) and $n = 4$ (right)

As shown in Figure 3, for $n = 3$ graphical evidence points to the fact that the approximate D-optimal design is obtained for $x_2 \to 0$ (or, equivalently, $x_2 \to 1$), which coincides with the optimal strategy for the estimation of θ; the same behaviour has been observed for $n = 4$. Nevertheless, numerical evidence shows that as n grows the D-optimal design depends upon the value of the correlation parameter.

As previously done, we now compare the locally D-optimal design ξ_θ^D with the space filling procedure in terms of efficiency:

$$D\text{-Eff} = \left(\frac{\det I(\beta, \theta; \bar{\xi})}{\det I(\beta, \theta; \xi_\theta^D)} \right)^{\frac{1}{3}} .$$

As shown in Figure 4, assuming the D-optimality criterion, the loss of efficiency induced by the space filling design is crucial, since the component of the Fisher information associated with the correlation parameter has a great impact on the D-criterion, particularly for small sample sizes.

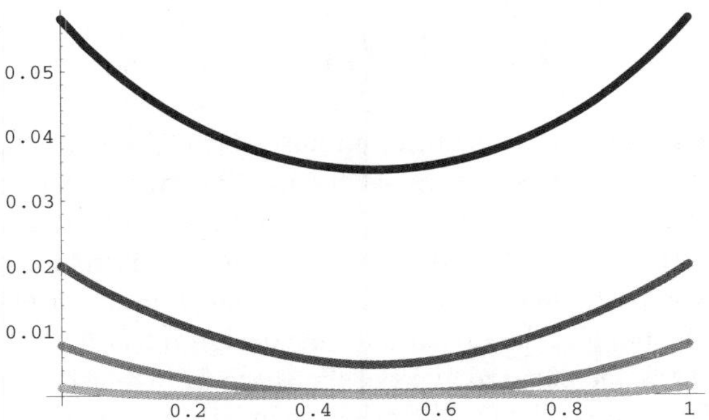

Fig. 3. D-optimality criterion: x_2 varies on the x-axis. The values of θ from the top to the bottom curve are 3, 5, 8, 20, respectively

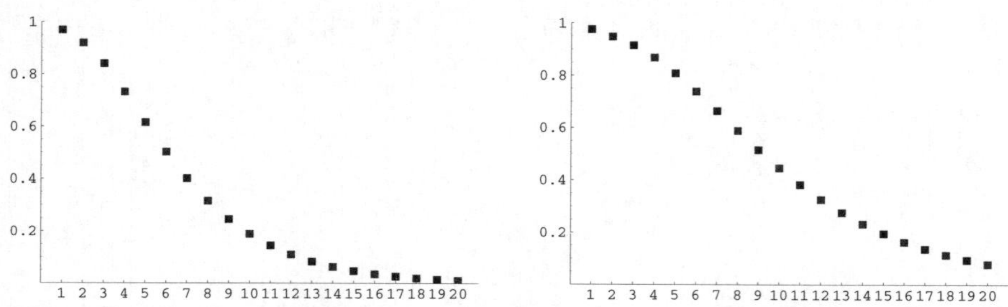

Fig. 4. Plots of the D-efficiency as θ varies with $n = 3$ (left) and $n = 4$ (right)

4.2 Optimal designs for prediction

As is well-known, accurate prediction represents one of the most important issues for computer experiments and therefore there is a need to introduce suitable criteria for choosing designs that predict the response well at untried points. Usually, this task is treated by minimising some functionals of the Mean Squared Prediction Error

$$E\left(\widehat{Y}(x) - Y(x)\right)^2 = 1 - (f^t(x); r^t(x)) \begin{pmatrix} 0 & F^t \\ F & C \end{pmatrix}^{-1} \begin{pmatrix} f(x) \\ r(x) \end{pmatrix}, \tag{9}$$

where $r(x)$ is the $(n \times 1)$ vector of correlations between the input x and the design points $x_1, \ldots, x_n$. In this framework, a very popular design criterion is the so-called *Integrated Mean Squared Prediction Error* (IMSPE), namely an average version of the MSPE in (9), given by

$$\begin{aligned} \text{IMSPE}(\xi) &= \int_{\Xi} \left[E\left(\widehat{Y}(x) - Y(x)\right)^2 \right] dx \\ &= 1 - tr\left[\begin{pmatrix} 0 & F^t \\ F & C \end{pmatrix}^{-1} \int_{\Xi} \begin{pmatrix} f(x)f^t(x) & f(x)r^t(x) \\ r(x)f^t(x) & r(x)r^t(x) \end{pmatrix} dx \right]. \end{aligned} \tag{10}$$

A design is called IMSPE optimal if it minimises (10). Under the Universal Kriging with exponential correlation structure (4), the IMSPE optimal design depends on the correlation parameter θ, and will be indicated by ξ_θ^P. For instance, when $n = 3$, as shown in Figures 5 and 6, the space filling design is optimal if the correlation is medium/high; whereas when the observations tend to be uncorrelated, the IMSPE criterion forces the point at the extremes of the design interval.

In order to check the performances of the space filling design for prediction, we have considered the ratio of the IMSPE criterion in (10) evaluated at the corresponding locally optimal design ξ_θ^P to the IMSPE($\bar{\xi}$), i.e.

$$\text{IMSPE}(\xi_\theta^P)/\text{IMSPE}\left(\bar{\xi}\right). \tag{11}$$

As emerged from Figure 7, for small sample sizes, space filling designs represent an approximately optimal strategy for prediction, since they are optimal for strongly

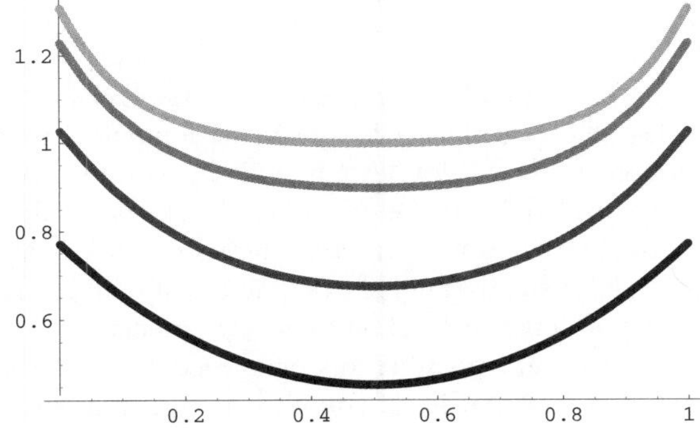

Fig. 5. IMSPE criterion as x_2 varies. The values of θ from the bottom curve to the top curve are 3, 5, 8, 10

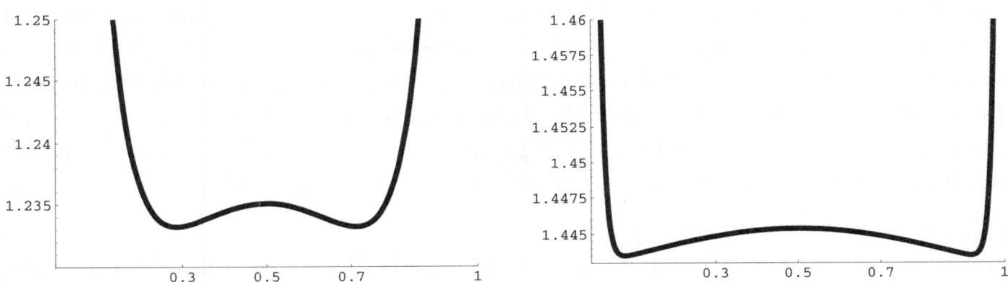

Fig. 6. IMSPE criterion as x_2 varies for $\theta = 20$ (left) and $\theta = 100$ (right)

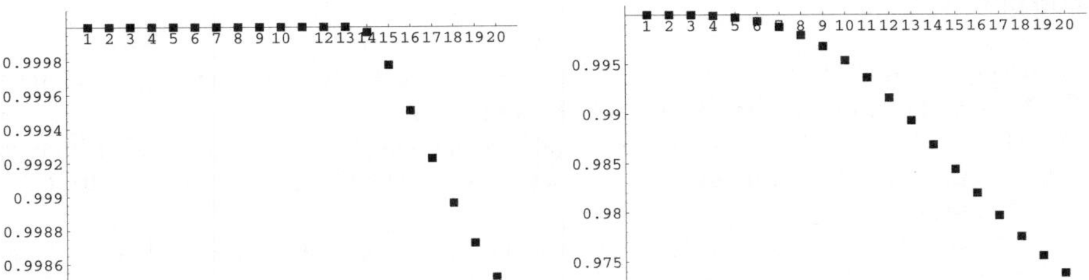

Fig. 7. Plots of the ratio in (11) as θ varies, with $n = 3$ (left) and $n = 4$ (right)

correlated observations and, at the same time, guarantee high efficiency as θ increases. Thus, by covering the full range of the input domain, space filling designs lead to good prediction over the entire experimental region.

5 Conclusions

This paper deals with the optimal design problem for the Universal Kriging model, emphasising the role of the space filling designs in the computer experiments field. In particular, the adoption of this class of designs could be harmful when interest lies in the estimation of the correlation parameter, while the equispaced design performs in an optimal manner, or almost optimal, in several situations related to different criteria. This is particularly true for small sample sizes in terms of both information gain and prediction, even in the presence of local optimality problems.

In the latter case, when it is possible to simulate extensively from the computer code a possible solution may consist of adopting a sequential approach which, by employing at each stage the observed outcomes for estimating the unknown parameters, can modify the design as it proceeds in order to converge to the unknown local optimum. A vast amount of literature on sequential designs for computer experiments based on the Kriging methodology is available. Usually, these methods consist of starting with a pilot study (e.g. a space filling design) in order to get an initial estimation of the unknown parameters of the model, and then a sequential procedure is activated via the adoption of a suitable allocation rule, which specifies at each step the optimal choice of the next design point. Finally, these procedures are stopped according to the specification of a stopping rule based on cost/time considerations or required amount of inferential accuracy. For a thorough description see [6, 8, 13, 19, 20].

Acknowledgement. This research was partly supported by the Italian Ministry for Research (MIUR) PRIN 2007 \Statistical methods for learning in clinical research.

References

1. Abrahamsen, P.: A review of Gaussian random fields and correlation functions. Technical Report 917, Norwegian Computing Center, Oslo (1997)
2. Abt, M., Welch, W.: Fisher information and maximum likelihood estimation of covariance parameter in Gaussian stochastic processes. The Canadian Journal of Statistics **26**, 127–137 (1998)
3. Atkinson, A., Donev, A.: Optimum experimental design. Clarendon Press, Oxford (1992)
4. Baldi Antognini, A., Zagoraiou, M.: Exact optimal designs for computer experiments via Kriging metamodelling. Journal of Statistical Planning and Inference, to appear (2010)
5. Bursztyn, D., Steinberg, D.M.: Comparisons of designs for computer experiments. Journal of Statistical Planning and Inference **136**, 1103–1119 (2006)
6. Gupta, A., Ding, Y., Xu, L.: Optimal Parameter Selection for Electronic Packaging Using Sequential Computer Simulations. Journal of Manufacturing Science and Engineering **128**, 705–715 (2006)
7. Kiseľák, J., Stehlík, M.: Equidistant and D-optimal designs for parameters of Ornstein-Uhlenbeck process. Statistics and Probability Letters **78**, 1388–1396 (2008)
8. Kleijnen, J.P.C., van Beers, W.C.M.: Application-driven sequential designs for simulation experiments: Kriging metamodelling. Journal of the Operational Research Society **55**, 876–883 (2004)

9. Lin, C.D., Bingham, D., Sitter, R.R., Tang, B.: A New and Flexible Method for Constructing Designs for Computer Experiments. Annals of Statistics, to appear (2009)
10. Müller, W.G., Stehlík, M.: Issues in the Optimal Design of Computer Simulation Experiments. Applied Stochastic Models in Business and Industry **25**, 163–177 (2009)
11. Pázman, A.: Criteria for optimal design of small-sample experiments with correlated observations. Kybernetika **43**, 453–462 (2007)
12. Pepelyshev, A.: Optimal Designs for the Exponential Model with Correlated Observations. In: López-Fidalgo, J., Rodríguez-Diáz, J.M., Torsney, B. (eds.) *mODa 8-Advances in Model-Oriented Design and Analysis*. Physica-Verlag, Heidelberg, 165–172 (2007)
13. Romano, D.: Sequential Experiments for Technological Applications: Some Examples. *Proceedings of XLIII Scientific Meeting of the Italian Statistical Society – Invited Session: Adaptive Experiments, Turin, 14–16 June 2006*, 391–402 (2006)
14. Sacks, J., Welch, W.J., Mitchell, T.J., Wynn, H.P.: Design and analysis of computer experiments. Statistical Science **4**, 409–423 (1989)
15. Santner, J.T., Williams, B.J., Notz, W.J.: The Design and Analysis of Computer Experiments. Springer, New York (2003)
16. Shewry, M.C., Wynn, H.P.: Maximum entropy sampling. Journal of Applied Statistics **14**, 165–170 (1987)
17. Stehlík, M.: D-optimal Designs and Equidistant Designs for Stationary Processes. In: López-Fidalgo, J., Rodríguez-Diáz, J.M., Torsney, B. (eds.) *mODa 8-Advances in Model-Oriented Design and Analysis*. Physica-Verlag, Heidelberg, 205–212 (2007)
18. Stehlík, M., Rodríguez-Díaz, J.M., Müller, W.G., López-Fidalgo, J.: Optimal allocation of bioassays in the case of parametrized covariance functions: an application to Lung's retention of radioactive particles. Test **17**, 56–68 (2008)
19. Van Beers, W.C.M., Kleijnen, J.P.C.: Kriging for interpolation in random simulation. Journal of the Operational Research Society **54**, 255–262 (2003)
20. Williams, B.J., Santner, J.T., Notz, W.I.: Sequential designs of computer experiments to minimize integrated response functions. Statistica Sinica **10**, 1133–1152 (2000)
21. Ying, Z.: Maximum likelihood estimation of parameters under a spatial sampling scheme. The Annals of Statistics **21**, 1567–1590 (1993)
22. Zagoraiou, M., Baldi Antognini, A.: Optimal designs for parameter estimation of the Ornstein-Uhlenbeck process. Applied Stochastic Models in Business and Industry **25**, 583–600 (2009)
23. Zhu, Z., Zhang, H.: Spatial sampling design under the infill asymptotic framework. Environmetrics **17**, 323–337 (2006)
24. Zimmerman, D.L.: Optimal network for spatial prediction, covariance parameter estimation, and empirical prediction. Environmetrics **17**, 635–652 (2006)

Exploitation, integration and statistical analysis of the Public Health Database and STEMI Archive in the Lombardia region

Pietro Barbieri, Niccolò Grieco, Francesca Ieva, Anna Maria Paganoni and Piercesare Secchi

Abstract. We describe the nature and aims of the Strategic Program "Exploitation, integration and study of current and future health databases in Lombardia for Acute Myocardial Infarction". The main goal of the Programme is the construction and statistical analysis of data coming from the integration of complex clinical and administrative databases concerning patients with Acute Coronary Syndromes treated in the Lombardia region. Clinical data sets arise from observational studies about specific diseases, while administrative data arise from standardised and on-going procedures of data collection. The linkage between clinical and administrative databases enables the Lombardia region to create an efficient global system for collecting and storing integrated longitudinal data, to check them, to guarantee their quality and to study them from a statistical perspective.

Key words: health service research, biostatistics, data mining, generalised linear mixed models

1 Introduction

The major objective of the two year Strategic Program "Exploitation, integration and study of current and future health databases in Lombardia for Acute Myocardial Infarction" (IMA Project), funded by the Ministry of Health and by "Direzione Generale Sanità – Regione Lombardia", and started in January 2009, is the identification of new diagnostic, therapeutic and organisational strategies to be applied to patients with Acute Coronary Syndromes (ACS), in order to improve clinical outcomes.

The experience of the Milan network for cardiac emergency shows how a networking strategy that coordinates territory, rescue units and hospitals in a complex urban area, with high technological and medical resources, improves the health care of patients with ST-segment Elevation Myocardial Infarction (STEMI) and provides the opportunity to collect and analyse data in order to optimise resources. Between 2006–2009 a pioneer pilot study called MOMI^2 (MOnth MOnitoring Myocardial Infarction in MIlan) was conducted by The Working Group for Cardiac Emergency in Milan, the Cardiology Society, and the 118 Dispatch Centre (national free number for medical emergencies) in the urban area of Milan to streamline an optimal care process for patients with STEMI. The statistical analyses of data collected during

Mantovan, P., Secchi, P. (Eds.): Complex Data Modeling and Computationally Intensive Statistical Methods

six time periods lasting from 30 to 60 days (MOMI^2.1–MOMI^2.6) have supported best clinical practice which states that an early pre-alarm of the Emergency Room (ER) is an essential step to improve the clinical treatment of patients. Pre-hospital and in-hospital times have been highlighted as fundamental factors we can act on to reduce the in-hospital mortality and to increase the rate of effective reperfusion treatments of infarcted related arteries. In particular the study proved that, in order to make the Door to Balloon time (DB)[1] lower than 90 minutes – a limit suggested by the American Heart Association/American College of Cardiology (AHA/ACC) guidelines – it is fundamental to take and transmit the electrocardiogram as soon as possible.

The results of these pilot studies indicated that a structured and efficient network of transport (118) and hospitals makes the difference in achieving best clinical results. This has driven the Lombardia region to design a wider plan, starting from the MOMI^2 experience, in order to construct an archive concerning patients with ACS and involving all the cardiology divisions of hospitals in Lombardia. The innovative idea in this project is not only to guarantee the same procedures to such an extended and intensive care area, but also to integrate data collected during this observational study with administrative databases (Public Health Databases – PHD) arising from standardised and on-going procedures of data collection; up to now these PHD have been used only for monitoring and managing territorial policies.

The innovative result of this plan is then the construction for each patient of an integrated longitudinal data vector containing both clinical histories and follow-ups on which advanced statistical analysis can be performed. These analyses are of paramount interest. Indeed, information coming from such data is much more informative and complete than those coming separately from clinical registers or administrative databases. For instance, we now have access to information concerning related pathologies and repeated procedures.

Other different experiences encouraged the effort of the Lombardia region in designing and supporting this challenging project. In particular it is worth mentioning the experience of the Strategic Programme: "Detection, characterisation and prevention of Major Adverse Cardiac Events after Drug Eluting Stent implantation in patients with Acute Coronary Syndrome", developed in the Emilia Romagna region, and the implementation of the REAL registry (*REgistro regionale AngiopLastiche dell'Emilia-Romagna*). This is a large prospective web-based multicentre registry designed to collect clinical and angiographic data of all consecutive Percutaneous Coronary Interventions (PCI) performed in a four million resident Italian region. Thirteen public and private centres of interventional cardiology participate in data collection. Procedural data are retrieved directly and continuously from the resident databases of each laboratory, which share a common pre-specified dataset. In this case, follow-up is obtained directly and independently from the Emilia-Romagna Regional Health Agency through the analysis of the hospital discharge records and mortality

[1] Door to Ballon time is a time measurement in emergency cardiac care. The interval starts with the patient's arrival in the emergency department, and ends when a catheter guide wire crosses the culprit lesion in the cardiac cath lab.

registries. This ensures a complete follow-up for 100% of patients resident in the region [33,34]. The existence of this parallel register proves the scientific interest and relevancy of these procedures in health care policy.

The relevance of the project is also proved by the fact that in the future it will change data collection making it a standardised and compulsory procedure for all hospitals in Lombardia[2] [8].

In the following section we present the experience of the MOMI^2 survey and the statistical analyses performed on it. This seminal experience in the urban area of Milan has been the motivating stimulus for the wider Strategic Programme. We then illustrate the new register designed for the Strategic Programme (Section 3), called the STEMI Archive, and the administrative data banks available for data integration (Section 4). Finally, in the last section, we describe the statistical techniques that will be applied for analysing data generated by the patients involved in the study (Section 5).

2 The MOMI^2 study

A net connecting the territory of 23 hospitals by a centralised coordination of emergency resources has been activated in the urban area of Milan since 2001. Its primary aims are promoting the best utilisation of different reperfusion strategies, reducing transport and decisional delays connected with logistics and therapies, and increasing the number of patients undergoing primary Percutaneous Coronary Intervention (PCI) within 90 minutes of arrival at the Emergency Room. Difficulties in reaching these goals are primary due to the fact that the urban area of Milan is a complex territory with a high density population (2.9 million resident and 1 million commuters daily) and 27 hospitals, and a great number of different health care structures. Twenty-three of them have a Cardiology Division and a Critical Care Unit; 18 offer a 24 hour available Cath Lab for primary PCI, and 5 are completed with a Cardiac Surgery unit. In order to monitor network activity, time to treatment and clinical outcomes, data collected for MOMI^2 related to patients admitted to hospitals belonging to the net was planned and made during six periods corresponding to six monthly/bimestral collections (respectively: MOMI^2.1 from Jun 1st to 30th 2006, MOMI^2.2 from Nov 15th to Dec 15th 2006, MOMI^2.3 from Jun 1st to Jul 30th 2007, MOMI^2.4 from Nov 15th to Dec 15th 2007, MOMI^2.5 from Jun 1st to 30th 2008, MOMI^2.6 from Jan 28th to Feb 28th 2009). The whole dataset collects data relative to 841 patients.

The experience of the Milan network for cardiac emergency shows how a networking strategy that coordinates territory, rescue units and hospitals in a complex urban area with high technological and medical resources, improves health care of patients with STEMI and provides the opportunity to collect and analyse data in order to optimise resources. There was a great number of patients treated with reperfusion therapy (82%) with a low hospital mortality (6.7%), an extensive use of PCI (73%), and a continuous attempt to reduce DB time. Almost 62% of patients met the guideline recommendations with a DB time of less than 90 minutes.

[2] Standardised and compulsory procedures for collecting data and sending them to Lombardia data banks are called *Debito Informativo*.

The analysis of the data collected in the MOMI^2 surveys show that (see [13, 17, 18]) the DB time is greatly influenced by organisational pre-hospital and in-hospital factors. In particular, we found that timing of the first ECG, means of transport to hospital, pre-alert, direct fast track to the Cath Lab and presentation at the hospital during working hours, were all relevant factors for the prediction of a DB time of less than 90 minutes. Of particular interest was the finding that execution and transmission of pre-hospital ECG (23% of patients), as well as triage within 10 minutes from ER presentation (59% of patients), were the two most important predictive factors in reducing DB time.

The analysis also focused on the dependence of principal performance indicators (in particular times to treatment) and clinical outcomes. Other studies have found conflicting results regarding the relationship between mortality and time to reperfusion with PCI. Some investigators have found a lower mortality for shorter symptom onset-to-balloon times either for all patients, or just certain subgroups such as those at high-risk [6]. Other studies did not find a lower mortality for shorter symptom onset-to-balloon times, but did find a lower mortality for shorter DB time [23]. Finally, some studies failed to find an association between mortality and pre-hospital and in-hospital times [27]. We detected [18] a connection between outcomes and times (both concerning symptom onset and in-hospital times); in particular our analysis pointed out the dependence between the efficacy of the reperfusion therapy (measured as a 70% reduction of the ST-segment elevation an hour after the PCI) and DB time and symptom onset time.

Details of the analyses are given in [18]. Dependence between the DB time and factors we can act upon in order to reduce it, has been explored by means of CARTs [4]. Indeed a CART analysis using Gini's impurity index splits groups satisfying or not the limit of 90 min for DB time in terms of time of first ECG within 10 minutes or not (see also [36]), limits suggested by the AHA/ACC guidelines. In fact, the distribution of the DB time in the population of patients with the first ECG within 10 minutes is confirmed to be stochastically smaller than the corresponding distribution in patients with the first ECG after 10 minutes; this stochastic order between the two distributions is confirmed by a Mann-Whitney test: p-value $< 10^{-12}$. In order to assess the discriminatory power of covariates we performed a random forest analysis [5] applied to CART predictors. The length of bars in the right panel of Figure 1 is proportional to the discriminatory power of each variable in splitting the groups of patients satisfying or not the limit of 90 minutes for DB time. Time of first ECG and way of admittance are pointed out as the most important covariates to distinguish the two groups. Investigation on the dependency structure between these two covariates showed a masking effect between the covariates detected by the classification analysis. An exact Fisher test, performed on the contingency table of the mode of hospital admittance and a variable indicating if the time of first ECG is within or not 10 minutes, shows strong statistical evidence (p-value $< 10^{-10}$) of dependence between these two covariates. The message generated by these analyses is always the same: to have a DB time lower than 90 minutes, it is fundamental to take and transmit an ECG as soon as possible.

The main outcomes (in-hospital survival and reperfusion efficacy) have been described through linear logistic regressions as functions of the other variables in the

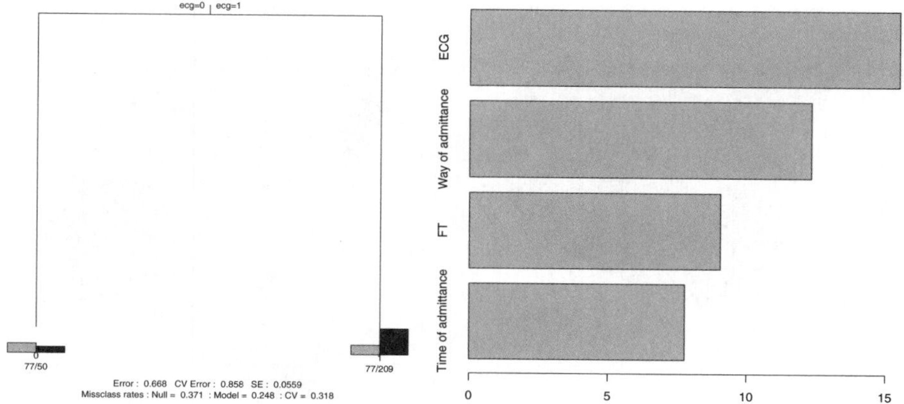

Fig. 1. Left panel: CART analysis: the groups satisfying (right path) or not (left path) the limit of 90 minutes for DB time are split in terms of first ECG within 10 minutes or not. Right panel: Random Forest on CART predictors (drawn in the left panel) assessing discriminatory power of covariates

dataset, in order to explore the relevance of the covariates in the improvement of the two performance indicators. In-hospital survival and reperfusion efficiency are both binary independent variables (Fisher exact test: p-value = 0.244). Denoting the outcome variable under study as Y, and the set of p predictors as $\mathbf{X}$, a linear logistic regression model for the binary response Y can be written as

$$\mathrm{logit}\{\mathrm{P}(\mathbf{X})\} \equiv \log\left\{\frac{\mathrm{P}(\mathbf{X})}{1-\mathrm{P}(\mathbf{X})}\right\} = \alpha + \sum_{\mathrm{j}=1}^{\mathrm{p}} \beta_{\mathrm{j}} \mathrm{X}_{\mathrm{j}}\,, \tag{1}$$

where $P(\mathbf{X}) = \mathrm{pr}(Y = 1|\mathbf{X})$. Clinical best practice and a stepwise model selection procedure based on backward selection with AIC criterion, pointed out killip class, age and total ischemic time (symptom onset to balloon time) as explanatory variables for the survival outcome. On the other hand, DB time and Symptom onset time have been selected as explanatory variables for reperfusion efficacy. Figures 2 and 3 illustrate the results; the surfaces describe the probability of in-hospital survival and of effective reperfusion, respectively. In-hospital survival probability decreases with increasing age and total ischemic time, for both the cases of less and more severe STEMI (measured by the killip class), but more strongly in the latter case. The probability of effective reperfusion decreases with increasing symptom onset and DB time.

These results support the effort of acting on some control variables (such as the reduction of DB time) in order to obtain an improvement in performance indicators and thus to increase the probability of successful treatment. The significance of symptom onset time in modeling the reperfusion efficacy suggests that it would be very important to persuade the population to call the free emergency number as soon as possible after symptom onset.

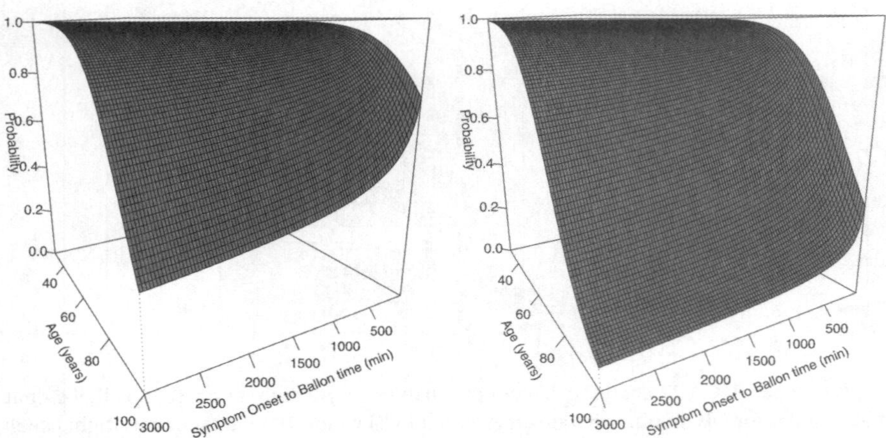

Fig. 2. Left panel: In-hospital survival surface estimated by a linear logistic regression model (killip class = 1 or 2). Right panel: In-hospital survival surface estimated by a linear logistic regression model (killip class = 3 or 4)

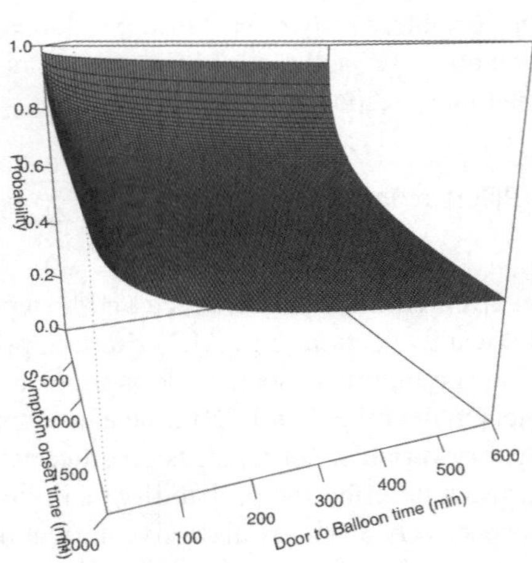

Fig. 3. Reperfusion efficacy surface estimated by a linear logistic regression model

3 The STEMI Archive

In this section we describe the aims and contents of the STEMI Archive. The archive has been designed according to the aims of the Strategic Programme and will be ready to be tested by the end of 2009. It consists of the collection of clinical information related to all patients admitted to hospitals of the Lombardia Network with STEMI diagnosis, as in the MOMI^2 collection for the urban area of Milan. From the information contained in the archive it will be possible to construct a data set where each patient will be represented by a profile with the following entries: individual serial number, date of birth, sex, time of symptoms onset, type of symptoms, time to call emergency services, type of rescue unit sent (advance or basic rescue unit, that is with or without pre-hospital 12d ECG teletransmission), site of infarction on ECG, mode of hospital admission, blood pressure and cardiac frequency at presentation, history of cardiac pathology, pre-hospital medication, date and hour of angioplasty (wiring), culprit lesion, ST resolution at 60 minutes, MACE (Major Adverse Cardiovascular Events) and Ejection Fraction at discharge. Personal data is collected so that the patient can be identified and a complete follow-up can be recorded. Other data are reported to evaluate critical times (symptom onset time, door to balloon time, door to needle time, time of first ECG and first medical contact to reperfusion time) with the aim of designing a preferential therapeutic path to reperfusion in STEMI patients, and to direct patient flow through different pathways according to time (e.g.: on hours vs off hours) or clinical conditions (killip class 1 or 2 vs others). Finally, information concerning results and outcomes of the procedures will be resumed in records attesting to whether a subject survives or not, and if the reperfusive procedure was effective or not. As in the MOMI^2 study, these data will represent some of the principal outcomes of interest. The STEMI Archive should overcome the difficulties faced with MOMI^2 data collections related to non-uniformity, inaccuracy of filling and data redundance. In particular, non-uniformity of data collection among different structures, or among successive surveys, and inaccuracy in filling data set fields will cease to be a problem because the archive procedure for collecting data will become mandatory for all hospitals through a directive issued by the Lawmaker [24, 26]. All centres will fill in the registry using the same protocol and with the same software; all fields of the data bank have been agreed by opinion leaders and scientific societies of cardiologists and a unique data collector was identified in the Governance Agency for Health, that will also be the data owner.

4 The Public Health Database

In this section we describe the structure, aim and use of the Lombardia Public Health Database. Up to now, this database has been used for administrative purposes only, since decision makers of health care organisations need information about the efficacy and costs of health services. Randomised Controlled Trials (RCTs) remain the accepted "gold standard" for determining the efficacy of new drugs or medical procedures. Randomised trials alone, however, cannot provide all the relevant information that decision makers need to weigh the implications of particular policies which affect medical therapies. Moreover, quality organisations and professional societies need information about the applicability of trial findings to the settings and patients of interest. Research using disease and intervention registries, outcome studies using administrative databases and performance indicators adopted by quality improvement methods can all shed light on who is most likely to benefit, what the important trade-offs are and how policy makers might promote the safe, effective and appropriate use of new interventions.

4.1 Healthcare databases

Administrative health care databases today play a central role in epidemiological evaluation of the Lombardia health care system because of their widespread diffusion and low cost of information. Public health care regulatory organisations can assist decision makers in providing information based on available electronic health records, promoting the development and the implementation of the methodological tools suitable for the analysis of administrative databases and answering questions concerning disease management. The aim of this kind of evaluation is to estimate adherence to best practice (in the setting of evidence based medicine) and the potential benefits and harms of specific health policies. Health care databases can be analysed in order to calculate measures of quality of care (quality indicators); moreover the implementation of disease and intervention registries based on administrative databases could enable decision makers to monitor the diffusion of new procedures or the effects of health policy interventions.

The Lombardia Region Data Warehouse, called "BDA" (*Banca Dati Assistito*), contains a huge amount of data and requires specific and advanced tools and structures for data mining and data analysis. The structure adopted by the Lombardia region is called Star Scheme [22], since it is centred on three main databases (*Ambulatoriale, Farmaceutica, Ricoveri*) containing information about visits, drugs, hospitalisations, and surgical procedures that took place in hospitals in Lombardia, while being supported by secondary databases (*Assistibili, Medici, Strutture e Farmacie, Farmaci, Codici Diagnosi e Procedure Chirurgiche*) which contain specific information about procedure coding or anagraphical information about people involved in the care process. The Star Scheme does not allow for repetition in record entering, i.e. just one record for each person is allowed. Records may be linked in order to achieve the correct information about the basic observation unit (i.e. the individual patient/subject). However, each of the above databases has its own dimension and structure, and data are different and differently recorded from one database to another. Suitable techniques are therefore required to make information coming from different databases uniform. The longitudinal data that we will analyse will be generated by deterministic record linkage tools between the STEMI Archive and the *Ambulatoriale* and *Ricoveri* databases of the BDA, and by probabilistic record matching [10] between the STEMI Archive and *Farmaceutica* database which is not entirely based on the patient's national insurance number (*Codice Fiscale*).

There is an increasing agreement among epidemiologists on the validity of disease and intervention registries based on administrative databases [2,3,9,14,25,38]. This motivated the Lombardia region to use its own administrative database for clinical and epidemiological aims. The most critical issue when using administrative databases for observational studies is represented by the selection criteria of the observation units: several different criteria may be used, and they will result in different images of prevalence or incidence of disease. Statistical analysis can be performed by means of multiple logistic regression models for studying outcomes and by means of survival analysis when studying failure times (hospital readmissions, continuity of drug prescriptions, survival times). Multilevel models can also be adopted if structural and organisational variables are measured. When outcomes are the main focus of the observational study, appropriate risk adjustment tools are needed. Hospital discharge records may be analysed with the indicators developed by the Agency for Health Care Research and Quality (AHRQ) that include efficient risk adjustment tools within a multiple logistic regression model. In disease management programmes the Johns Hopkins Adjusted Clinical Groups (ACG) methodology and the Classification Reasearch Group (CRG) classification system have been proposed [11,16,35].

4.2 Health information systems in Lombardia

Health information systems in Lombardia experienced a rapid growth as a consequence of the introduction to Italian health management of Diagnosis Related Groups (DRGs) in 1995. The development of health care measures for the specific aim of health system financing, gave rise to the availability of information useful for evaluating the efficiency of the providers and the efficacy of their activities. The development of health information systems was particularly pronounced in hospitals, and this extended the possibilities of measuring their activities: from the "classic" indicators (average length of stay, occupancy rate, turnover interval) measuring bare hospitality, to more meaningful evaluations linked to patient classification systems and to the actual opportunity of calculating quality indicators. Several regional and national rules introduced in recent years a large number of indicators in the Italian national health system. However, most of these indicators measure only a few aspects of the health system: costs, degree and characteristics of supply, organisational factors, access to health care, population health status [21]. Few indicators measure patient outcomes or evaluate the processes within the hospitals; indications about criteria for the definition of such measures are scanty and research about the validation of the indicators has not been properly developed. On the basis of these considerations the National Agency for Regional Health Services in Lombardia (*Agenzia per i Servizi Sanitari Regionali* – ASSR) developed a set of quality measures (outcome and process indicators) in the context of the Strategic Programme founded by the Ministry of Health.

Indeed, one of the main goals of the Strategic Programme is finding a set of indicators useful for comparison of health care providers and for the identification of factors which can produce different outcomes. Interpretation of the results will need to be supported by information collected to measure confounding factors due to differences in case-mix or selective health care options. This will be obtained using the STEMI Archive.

5 The statistical perspective

In this section we describe the statistical tools that we will use in order to model in-hospital survival and treatment efficacy outcomes. The identification of principal prognostic factors of outcomes is the main goal of the statistical analysis we will conduct. Some preliminary results obtained in a pilot study [19] support the choice of methodological approach we will use in the analysis of data available after the linkage between the STEMI Archive and hospital discharge data from the Public Health Database. Results from these preliminary statistical analyses are very promising; unfortunately we are not yet permitted to discuss them in public, because the matter is covered by a non-disclosure agreement with the health governance of the Lombardia region.

We identified suitable statistical techniques to make an effective dimensional reduction of the complex longitudinal data vectors representing patients. Frailty models appear to be appropriate for capturing and characterising information arising from hospital discharge data. Data coming from health databases are usually affected by huge variability, called overdispersion. The main cause of this phenomenon is the grouped nature of data: each patient is a grouping factor with respect to its own admissions, while hospitals are a grouping factor with respect to admitted patients, and so on. In this study, we will model outcomes using the hospital as a grouping factor. The choice is based on clinical considerations and practical evidence. Indeed, a negative outcome can be due to bad performance of the structure with respect to a patient (without considering the initial conditions of the patient), but also to good performance with respect to a patient who arrived in a very bad condition. After splitting the effect on outcome due to the hospital, from the outcome variability due to the different patient initial conditions, we would be in a position to generate health indicators of performance, and a benchmark that will make hospitals aware of their standing in the wider regional context. These goals could be achieved by fitting generalised linear mixed models on data coming from the integrated database.

In the following subsections we summarise three different tools of complex data statistical modeling: frailty models, generalised linear mixed models and Bayesian hierarchical models. Frailty models have been widely and successfully used (see for instance [31, 37]) to model the hazard function of the patient hospitalisation process. The estimated hazard functions, different between patients because of the random effect of the latent frailty variable, could be used as a functional data to cluster different risk subpopulations, or as a prognostic factor for the outcomes of the acute event registered in the STEMI Archive. Generalised linear mixed models seem very promising to model and explain the binary or counting outcomes of interest, not only adjusting the analysis with respect to the traditional fixed covariates, but also taking into account the overdispersion due to the grouped nature of data. Some preliminary results concerning the use of mixed models on MOMI^2 data are summarised in [20]. Bayesian hierarchical models have been used to study variations in health care utilisation for multilevel clustered data, such as patients clustered by hospital and geographic origin (see for instance [7, 29]). The data collected in this project present exactly a multilevel clusterd structure.

5.1 Frailty models

Firstly we study the clinical history of patients, and in particular the sequence of hospital discharge data coming from the Public Health Database. We focus on a general class of semiparametric models for recurrent events, such as hospitalisations, proposed by Peña and Hollander [30, 31]. Consider a patient that is being monitored for the occurrence of a recurrent event over a time period $[0, \tau]$; τ could be a random time (for example the time registered in the STEMI Archive) following an unknown probability distribution function. Let $0 \equiv S_0 < S_1 < \cdots$ be the random times of occurrences. Let $\mathbf{X(s)}$ be a possibly time-varying, observable q-dimensional vector of covariates such as gender, age, concurrent diseases. So we are dealing with the trajectories of the following counting process:

$$N(s) = \sum_{j=1}^{+\infty} \mathbf{I}\{S_j \leq s,\ S_j \leq \tau\},$$

which represents the number of occurrences of the recurrent event (hospitalisation) during the period $[0, s]$; in [31] the authors propose a general model for the hazard rate function $\lambda(s)$ of the process N:

$$\lambda(s|Z, \mathbf{X}) = Z\lambda_0(s - S_{N(s^-)})\rho(N(s^-); \alpha)\psi(\beta^{\mathbf{t}}\mathbf{X}(s)). \tag{2}$$

Z is a random variable which represents the unobservable frailty of the patient, λ_0 is an unknown baseline hazard rate function, the function $\rho(\cdot; \alpha)$ incorporates the effect of the accumulating event occurrences and the link function ψ summarises the covariates contribution. Many authors [15, 37] interpret frailty as modeling the effect of an unobserved covariate which leads some patients to have more occurrences than others. In particular, (2) is a random effect model for time-to-event data where the random effect has a multiplicative effect on the baseline hazard function. By assuming specific forms for the law of Z some elegant mathematical results can be derived; in fact, a common choice for the unknown frailty distribution is a gamma distribution with unit mean and variance $1/\xi$ ($Z \sim \Gamma(\xi, \xi)$). Imposing the restriction that the gamma shape and scale parameters are equal guarantees model identifiability (see [31]) and estimation of model parameters $(\xi, \lambda_0(\cdot), \alpha, \beta)$.

5.2 Generalised linear mixed models

Generalised linear mixed models (GLMM) are a natural extension of Generalised Linear Models (GLM). GLM extend ordinary regression by allowing non-normal responses and a link function of the mean. GLMM is a further extension that permits random effects as well as fixed effects in the linear predictors. An extensive overview on these topics can be found in [1, 12, 32]. In this regression setting, parameters that describe factor effects in ordinary linear models are called fixed effects; they apply to all categories of interest. By contrast, random effects usually apply to a sample. For a study using a sample of hospitals the model treats observations from a given hospital as a cluster, and assumes a random effect for each hospital. Let Y_{ij} be the response of subject i in cluster $j, i = i_1, ..., i_j$. In our case the responses are not only in-hospital survival and reperfusion efficacy, but also number of re-hospitalisations or re-procedures after the trigger event registered in the STEMI Archive. The number of observations may vary by cluster. Let $\mathbf{X}_{ij}$ denote a vector of explanatory variables, such as age, procedure times, symptom onset times, estimated frailty, for fixed effect model parameters γ. Let $\mathbf{U}_j$ denote the vector of random effects for hospital j (for example exposure of the hospital). This is common to all observations in the cluster. Let $\mathbf{Z}_{ij}$ be a vector of their explanatory variables. Let $\mu_{ij} = \mathbb{E}(Y_{ij}|\mathbf{U}_j)$. The linear predictor for a GLMM has the form

$$g(\mu_{ij}) = \gamma^{\mathbf{t}}\mathbf{X} + \mathbf{U_j^t}\mathbf{Z_{ij}}\,,$$

where g is the link function. For binary outcomes the link function g is the logit link. The random effect vector $\mathbf{U_j}$ is assumed to have a multivariate Normal distribution $\mathcal{N}_q(\mathbf{0}, \Sigma)$. The covariance matrix Σ depends on unknown variance components and possibly also correlation parameters.

The main goal is the joint estimation of (γ, Σ). Parameters pertaining to the random effects may be also of interest as a useful summary of the degree of heterogeneity of the population. Indeed the ratio between two components of fixed effects γ is the ratio of partial derivative of the log odds with respect to the corresponding covariates, and could be thought of as a measure of the relative covariates' strength in modeling the outcome.

5.3 Bayesian hierarchical models

GLMM are appealing when treating grouped data coming from health databases, but there are some computational problems, connected with the estimation of regression parameters, when they are used with binary data such as the outcomes of interest in our analysis.

The hierarchical model formulation where the outcome Y_{ij} is modeled conditionally on random effects, which are in turn then modeled in an additional step, makes the Bayesian paradigm appealing for interpreting and fitting GLMMs. The complete likelihood is in this case

$$L\left(\gamma, \Sigma\right) = \prod_{j=1}^{N} \int \prod_{i=1}^{n_j} f_{ij}\left(y_{ij}|\mathbf{u}_j, \gamma\right) f\left(\mathbf{u}_j|\Sigma\right) d\mathbf{u}_j \tag{3}$$

where N is the number of hospitals and n_j is the number of patients in each hospital. The key problem in maximising (3) is the presence of N integrals over the q-dimensional random effect $\mathbf{u}_j$; in the case of Bernoulli outcomes no analytic expression is available for these integrals and numerical approximations are needed. The most common approach approximates the integrand using the Laplace method. A different approach represents data as a sum of a mean and an error term, and is obtained as a Taylor expansion around the fixed effect component (Penalised Quasi Likelihood – PQL) or around the sum of fixed and random components (Marginal Quasi Likelihood – MQL). Unfortunately both these methods produce biased estimates of model parameters when applied to binary or unbalanced data; this is exactly the case we are dealing with in our readings. To overcome this problem, we will consider a direct numerical approximation of the likelihood.

The Bayesian approach to GLMM fitting is also appealing because the distinction between fixed and random effects no longer occurs, since every effect has a probability distribution. Then, Markov Chain Monte Carlo (MCMC) methods can be used for approximating intractable posterior distributions and their mode [28].

6 Conclusions

Studies such as MOMI^2 surveys and previous experiences of data collection carried out in the Milan intensive care area, have been seminal for two projects with broader aimes. The first is the *Progetto PROMETEO* (PROgetto Milano Ecg Teletrasmessi ExtraOspedaliero), whose goal is to provide all basic rescue units operating in the urban area of Milan with the ECG teletrasmission equipment. Motivation for the project comes directly out of the evidence provided by MOMI^2 results on the fundamental role played by an early ECG in improving survival outcome of STEMI, and highlights how the effort of monitoring data from a statistical perspective has a deep social impact. The second project, which represents the main target of the Strategic Programme, aims at extending the MOMI^2 paradigm for collecting and analysing data to all cardiology divisions of hospitals operating in the Lombardia region. The creation of an efficient regional network to face the ST-segment Elevation Myocardial Infarction is made possible by the design of the STEMI Archive and its integration with the regional Public Health Database; the link between the two databases will generate the primary platform for the study of the impact and care of STEMI on the whole territory of the Lombardia region. Previous data gathering and statistical analysis restricted to the urban area of Milan were compelling for the realisation of this complex and challenging project. This innovative and pioneering experience should become a methodological prototype for the optimisation of health care processes in the Lombardia region.

Acknowledgement. This work is part of the Strategic Programme "Exploitation, integration and study of current and future health databases in Lombardia for Acute Myocardial Infarction" supported by "Ministero del Lavoro, della Salute e delle Politiche Sociali" and by "Direzione Generale Sanità – Regione Lombardia". The authors wish to thank the Working Group for Cardiac Emergency in Milan, the Cardiology Society, and the 118 Dispatch Centre.

References

1. Agresti, A.: Categorical Data Analysis. Wiley, New York (2002)
2. Balzi, D., Barchielli, A., Battistella, G. et al.: Stima della prevalenza della cardiopatia ischemica basata su dati sanitari correnti mediante un algoritmo comune in differenti aree italiane. Epidemiologia e Prevenzione **32**(3) 22–29 (2008)
3. Barendregt, J.J., Van Oortmarssen, J.G., Vos, T. et al.: A generic model for the assessment of disease epidemiology: the computational basis of DisMod II. Population Health Metrics **1**, (2003)
4. Breiman, L., Friedman, J.H., Olshen, R.A., Stone, C.J.: Classification and Regression Trees. Wadsworth & Brooks, Monterey, California (1984)
5. Breiman, L.: Random Forest. Machine Learning **45**(1), 5–32 (2001)
6. Cannon, C.P., Gibson, C.M., Lambrew, C.T., Shoultz, D.A., Levy, D., French, W.J., Gore, J.M., Weaver, W.D., Rogers, W.J., Tiefenbrunn, A.J.: Relationship of Symptom-Onset-to-Balloon Time and Door-to-Balloon Time with Mortality in Patients undergoing Angioplasty for Acute Myocardial Infarction. Journal of American Medical Association **283** (22), 2941–2947 (2000)

7. Daniels, M.J., Gastonis, C.: Hierarchical Generalized Linear Models in the Analysis of Variations in Health Care Utilization. Journal of the American Statistical Association **94** (445), 29–42 (1999)
8. Determinazioni in merito alla "Rete per il trattamento dei pazienti con Infarto Miocardico con tratto ST elevato(STEMI)": Decreto N° 10446, 15/10/2009, Direzione Generale Sanità – Regione Lombardia (2009)
9. Every, N.R., Frederick, P.D., Robinson, M. et al.: A Comparison of the National Registry of Myocardial Infarction With the Cooperative Cardiovascular Project. Journal of the American College of Cardiology **33**(7), 1887–1894 (1999)
10. Fellegi, I., Sunter, A.: A Theory for Record Linkage. Journal of the American Statistical Association **64**(328), 1183–1210 (1969)
11. Glance, L.G., Osler, T.M., Mukamel, D.B. et al.: Impact of the present-on-admission indicator on hospital quality measurement experience with the Agency for Healthcare Research and Qualità (AHRQ) Inpatient Quality Indicators. Medical Care **46**(2), 112–119 (2008)
12. Goldstein, H.: Multilevel Statistical Models. third edn., John Wiley & Sons, New York (2003)
13. Grieco, N., Corrada, E., Sesana, G., Fontana, G., Lombardi, F., Ieva, F., Paganoni, A.M., Marzegalli, M.: Le reti dell'emergenza in cardiologia : l'esperienza lombarda. Giornale Italiano di Cardiologia Supplemento "Crema Cardiologia 2008". Nuove Prospettive in Cardiologia **9**, 56–62 (2008)
14. Hanratty, R., Estacio, R.O., Dickinson L.M., et al.: Testing Electronic Algorithms to create Disease Registries in a Safety Net System. Journal of Health Care Poor Underserved **19** (2), 452–465 (2008)
15. Hougaard, P.: Life table methods for heterogeneous populations: Distributions describing the heterogeneity. Biometrika **71**, 75–83 (1984)
16. Hughes, J.S., Averill, R.F., Eisenhandler, J. et al.: Clinical Risk Groups (CRGs). A Classification System for Risk-Adjusted Capitation-Based Payment and Health Care Management. Medical Care **42**(1), 81–90 (2004)
17. Ieva, F.: Modelli statistici per lo studio dei tempi di intervento nell'infarto miocardico acuto. Master Thesis, Dipartimento di Matematica, Politecnico di Milano (2008) Available at: http://mox.polimi.it/it/progetti/pubblicazioni/tesi/ieva.pdf
18. Ieva, F., Paganoni, A.M.: A case study on treatment times in patients with ST-Segment Elevation Myocardial Infarction. MOX-Report, n. 05/2009, Dipartimento di Matematica, Politecnico di Milano (2009) Available at: http://mox.polimi.it/it/progetti/pubblicazioni/quaderni/05-2009.pdf
19. Ieva, F., Paganoni, A.M.: Statistical Analysis of an integrated Database concerning patients with Acute Coronary Syndromes. *S.Co.2009 – Sixth conference – Proceedings*. Maggioli, Milano, 223–228 (2009)
20. Ieva, F., Paganoni, A.M.: Multilevel models for clinical registers concerning STEMI patients in a complex urban reality: a statistical analysis of $MOMI^2$ 2 survey. MOX Report n. 08/2010, Dipartimento di Matematica, Politecnico di Milano (2010) Available at: http://mox.polimi.it/it/progetti/pubblicazioni/quaderni/08-2010.pdf. Submitted (2010)
21. Gli Indicatori per la qualità: strumenti, metodi, risultati. Supplemento al numero 15 di Monitor (2005) Available at: http://www.agenas.it/monitor_supplementi.html
22. Inmon, W.H.: Building the Data Warehouse. second edn., John Wiley & Sons, New York (1996)
23. Jneid, H., Fonarow, G.C., Cannon, C.P., Palacios, I.F., Kilic, T. et al.: Impact of Time of Presentation on the Care and Outcomes of Acute Myocardial Infarction. Circulation **117**, 2502–2509 (2008)

24. Krumholz, H.M., Anderson, J.L., Bachelder, B.L., Fesmire, F.M.: ACC/AHA 2008 Performance Measures for Adults With ST-Elevation and Non-ST-Elevation Myocardial Infarction. Circulation **118**, 2596–2648 (2008)
25. Manuel, D.G., Lim, J.J.Y., Tanuseputro, P. et al.: How many people have a myocardial infarction? Prevalence estimated using historical hospital data. BMC Public Health **7**, 174–89 (2007)
26. Masoudi, F.A., Bonow, R.O., Brindis, R.G., Cannon, C.P., DeBuhr, J., Fitzgerald, S., Heidenreich, P.A.: ACC/AHA 2008 Statement on Performance Measurement and Reperfusion Therapy. Circulation **118**, 2649–2661 (2008)
27. MacNamara, R.L., Wang, Y., Herrin, J., Curtis, J.P., Bradley, E.H. et al: Effect of Door-to-Balloon Time on Mortality in Patients with ST-Segment Elevation Myocardial Infarction. Journal of American College of Cardiology **47**, 2180–2186 (2006)
28. Molenberghs, G.,Verbeke, G.: Models for Discrete Longitudinal Data. Springer, New York (2000)
29. Normand, S-L. T., Shahian, D.M.: Statistical and Clinical Aspects of Hospital Outcomes Profiling. Statistical Science **22**(2), 206–226 (2007)
30. Peña, E., Hollander, M.: Models for recurrent events in reliability and survival analysis. In: Soyer, R., Mazzucchi, T. Singpurwalla, N. (eds.) *Mathematical Reliability: An Expository Perspective*. Kluwer Academic Publishers, Dordrecht, 105–123 (2004)
31. Peña, E., Slate, E.H., González, J.R.: Semiparametric inference for a general class of models for recurrent events. Journal of Statistical Planning and Inference **137**, 1727–1747 (2007)
32. Pinheiro, C, Bates, D.M.: Mixed-Effects Models in S and S-Plus. Springer, New York (2000).
33. Saia, F., Piovaccari, G., Manari, A., Guastaroba, P., Vignali, L., Varani, E., Santarelli, A., Benassi, A., Liso, A., Campo, G., Tondi, S., Tarantino, F., De Palma, R., Marzocchi, A.: Patient selection to enhance the long-term benefit of first generation drug-eluting stents for coronary revascularisation procedures. Insights from a large multicentre registry. EuroIntervention **5**(1), 57–66 (2009)
34. Saia, F., Marrozzini, C., Ortolani, P., Palmerini, T., Guastaroba, P., Cortesi, P., Pavesi, P.C., Gordini, G., Pancaldi, L.G., Taglieri, N., Di Pasquale, G., Branzi, A., Marzocchi A.: Optimisation of therapeutic strategies for ST-segment elevation acute myocardial infarction: the impact of a territorial network on reperfusion therapy and mortality. Heart **95**(5), 370–376 (2009)
35. Sibley, L.M., Moineddin, R., Agham,M.M. et al.: Risk Adjustment Using Administrative Data-Based and Survey-Derived Methods for Explaining Physician Utilization. Medical [Epub ahead of print] (2009)
36. Ting, H.H., Krumholtz, H.M., Bradley, E.H., Cone, D.C., Curtis, J.P. et al.: Implementation and Integration of Prehospital ECGs into System of Care for Acute Coronary Sindrome. Circulation (2008) Available at: http://circ.ahajournals.org
37. Vaupel, J.W., Manton, K.G., Stallard, E.: The Impact of Heterogeneity in Individual Frailty on the Dynamics of Mortality. Demography **16**, 439–454 (1979)
38. Wirehn, A.B., Karlsson, H.M., Cartensen J.M., et al.: Estimating Disease Prevalence using a population-based administrative healthcare database. Scandinavian Journal of Public Health **35**, 424–431 (2007)

Bootstrap algorithms for variance estimation in πPS sampling

Alessandro Barbiero and Fulvia Mecatti

Abstract. The problem of bootstrapping the estimator's variance under a probability proportional to size design is examined. Focusing on the Horvitz-Thompson estimator, three πPS-bootstrap algorithms are introduced with the purpose of both simplifying available procedures and of improving efficiency. Results from a simulation study using both natural and artificial data are presented in order to empirically investigate the properties of the provided bootstrap variance estimators.

Key words: auxiliary variable, efficiency, Horvitz-Thompson estimator, inclusion probability, non-iid sampling, probability proportional to size sampling, simulations

1 Introduction

In a complex survey sampling, every population unit $i \in U$ $(i = 1 \ldots N)$ is assigned a specific probability π_i to be included in the sample s. In addition, the random mechanism providing s usually violates the classical hypothesis of independent and identically distributed sample data (referred to as *iid* henceforth), for instance with clustering, multistage selection or without replacement selection. We assume the total $Y = \sum_{i \in U} y_i$ of a quantitative study variable y as the parameter to be estimated. We focus on a design without replacement and with inclusion probability proportional to an auxiliary variable x, *i.e.* setting $\pi_i \propto x_i/X$ where $X = \sum_{i \in U} x_i$ is the population auxiliary total. This is generally referred to as *IPPS* sampling or πPS sampling. Pairing a πPS sampling with the unbiased Horvitz-Thompson estimator $\hat{Y}_{HT} = \sum_{i \in s} y_i/\pi_i$ results in a strategy methodologically appealing since the variance of $\hat{Y}_{HT}$,

$$V(\hat{Y}_{HT}) = \sum_{i \in U} \frac{1-\pi_i}{\pi_i} y_i^2 + \sum \sum_{j \neq i \in U} \left(\frac{\pi_{ij}}{\pi_i \pi_j} - 1 \right) y_i y_j, \quad (1)$$

is reduced to zero as the relationship between x and y approaches proportionality.

From a practical point of view a variance estimator is essential for assessing the estimate's accuracy and for providing confidence intervals. For a fixed sample size n,

Mantovan, P., Secchi, P. (Eds.): Complex Data Modeling and Computationally Intensive Statistical Methods

the Sen-Yates-Grundy estimator [18,22] has the closed analytic form

$$v_{SYG} = \sum\sum_{i<j\in s} \frac{\pi_i\pi_j - \pi_{ij}}{\pi_{ij}} \left(\frac{y_i}{\pi_i} - \frac{y_j}{\pi_j}\right)^2, \quad (2)$$

and unbiased for strictly positive joint inclusion probabilities π_{ij} of pair of population units $i \neq j \in U$. However, v_{SYG} presents some drawbacks which might limit the applications. For instance, it is computationally unfeasible for involving a double summation, and for depending on the joint inclusion probabilities π_{ij} of pair of sampled units $i \neq j \in s$, which are usually too cumbersome to be exactly computed for a fixed sample size $n > 2$ [1,2,5]. In addition, v_{SYG} is not uniformly positive for any πPS design.

A bootstrap estimator, although numeric, is a natural alternative for addressing those issues for being positive by construction. It can be computed for any sample size and does not require the explicit knowledge of joint inclusion probabilities. Furthermore, since the bootstrap is a general tool, a πPS-bootstrap algorithm for variance estimation would apply to any estimator for any population parameter, no matter what its complexity. Thus, it is important to first explore the simple linear case where known analytical solutions are available for assessment purposes. In this paper some πPS-bootstrap algorithms are proposed with the main purpose of improving both the computational and the statistical efficiency of estimating the variance of the Horvitz-Thompson estimator $V(\hat{Y}_{HT})$ as given by Equation (1).

Since the original Efron's bootstrap applies in the classical *iid* setup [3,7], suitable modified bootstrap algorithms are needed in order to handle the complexity of the sampling design.

In Section 2 the original Efron's bootstrap is recalled and applied to sample data selected under a πPS design. In Section 3 a previous πPS bootstrap is presented as a starting point for the three modified algorithms proposed in Section 4 and 5. Empirical results from an extensive simulation study comparing 8 different variance estimators are given in Section 6, and some final remarks are given in Section 7.

2 The naïve boostrap

The Efron's bootstrap was originally developed in a classical *iid* framework. When directly applied in a more general context, for instance under a complex sampling design with no adaptation whatsoever, it is frequently referred to as *naïve boostrap*. The naïve boostrap estimator v_{naive} of $V(\hat{Y}_{HT})$ is provided by the following four-step standard algorithm:

- *resampling step*: from the observed sample s (henceforth named the *original sample*) an *iid* sample s^* of the same size n is drawn (henceforth named the *bootstrap sample*);
- *replication step*: on the bootstrap sample, the same Horvitz-Thompson estimator is computed resulting in the replication $\hat{Y}^* = \sum_{i^*\in s^*} y_{i^*}/\pi_{i^*}$ of the original estimator $\hat{Y}_{HT}$;

- *bootstrap distribution step*: the first two steps are iterated B times (B chosen to be sufficiently large) providing the bootstrap distribution $\left\{\hat{Y}_b^*, b = 1 \dots B\right\}$ of the original estimator $\hat{Y}_{HT}$;
- *variance estimation step*: the variance of the bootstrap distribution defines the bootstrap estimator of $V(\hat{Y}_{HT})$. In fact, in the present simple linear case, the bootstrap distribution step is not necessary since the variance of the replication $\hat{Y}^*$ can be analytically computed. Let ν_i be the random variable counting the frequency of every unit $i \in s$ included in the bootstrap sample s^*; it is straightforward that

$$E^*(\hat{Y}^*) = E^*(\sum_{i^* \in s^*} y_{i^*}/\pi_{i^*}) = E(\sum_{i \in s} y_i \nu_i/\pi_i) = \hat{Y}_{HT} \tag{3}$$

and

$$\begin{aligned} V^*(\hat{Y}^*) &= V\left(\sum_{i \in s} \frac{y_i}{\pi_i} \nu_i\right) = \sum_{i \in s} \left(\frac{y_i}{\pi_i}\right)^2 V(\nu_i) + \sum \sum_{i \neq j \in s} \frac{y_i}{\pi_i} \frac{y_j}{\pi_j} Cov(\nu_i, \nu_j) \\ &= \sum_{i \in s} \left(\frac{y_i}{\pi_i} - \frac{\hat{Y}_{HT}}{n}\right)^2 = \upsilon_{naive} , \end{aligned} \tag{4}$$

where * denotes expectation under the resampling design. Thus, except for the scaling coefficient $(n-1)/n$, υ_{naive} coincides with the customary with-replacement variance estimator of $V(\hat{Y}_{HT})$ (see for example [17]).

3 Holmberg's πPS bootstrap

Several proposals to adapt the original Efron's bootstrap to non-*iid* situations have already appeared in literature, mainly for without replacement selection. The *with-replacement* bootstrap [13] and the *without-replacement* bootstrap [3, 8] apply to a simple random sample (SRS) with equal inclusion probability $\pi_i = n/N$. The *rescaling* bootstrap [14] and the *mirror-match* bootstrap [20] address the *scaling problem* by using bootstrap samples of size $n^* \neq n$. Rao and Wu [14] also introduced a πPS bootstrap requiring in fact the computation of the joint inclusion probability π_{ij} as for the unbiased estimator υ_{SYG}.

The Gross-Chao-Lo SRS bootstrap is based on the concept of *bootstrap population*. Given the original sample s, the bootstrap population U^* is the pseudo-population formed by replicating N/n times every sampled unit $i \in s$. Hence, the bootstrap population includes data from the original sample only and has the same size as the original population $N^* = N$ according to a basic bootstrap principle [4]. A bootstrap sample of the same size n is then selected from U^* under an SRS design, *i.e.* by mimicking the original sampling design, still according to a basic bootstrap principle. As a consequence the bootstrap population approach proves to be methodologically appealing. However, computationally, since N/n may not be an integer, a further step performing the randomisation between the integer part $\lfloor N/n \rfloor$ and $\lfloor N/n \rfloor + 1$

must be added in order to deal with the general case. Holmberg [12] generalised the bootstrap population approach for πPS design by using the inverse of the inclusion probability $\pi_i = nx_i/X$ to construct U^*.

Let $1/\pi_i = c_i + r_i$ where $c_i = \lfloor \pi_i^{-1} \rfloor$ and $0 \leq r_i < 1$. Let ε_i be the realisation of n independent Bernoulli trials each with probability r_i. Finally define $d_i = c_i + \varepsilon_i$. The bootstrap variance estimator v_H of $V(\hat{Y}_{HT})$ is provided by the following original Holmberg's algorithm:

- *bootstrap population step*: construct $U^* = \{1^* \cdots i^* \cdots N^*\}$ replicating d_i times each sampled unit $i \in s$. Thus the bootstrap population size is given by $N^* = \sum_{i \in s} d_i$ and the bootstrap population auxiliary total by $X^* = \sum_{i \in s} d_i x_i$;
- *resampling step*: a bootstrap sample s^* is selected from U^* by mimicking the original sample, *i.e.* with the same size n and under the same πPS design, with (resampling) inclusion probabilities $\pi_{i^*} = nx_{i^*}/X^*$.

The remaining three step (replication, bootstrap distribution and the variance estimation step) are then performed as in the standard (naïve) bootstrap algorithm in Section 2, eventually giving the Holmberg variance estimator v_H as the variance of the bootstrap distribution

$$v_H = \frac{1}{B-1} \sum_{b=1}^{B} \left(\hat{Y}_b^* - \frac{1}{B} \sum_{b=1}^{B} \hat{Y}^* \right)^2 . \tag{5}$$

Notice that if $r_i = 0$ holds $\forall i \in s$, *i.e.* when π_i^{-1} are integers for all sampled units, then the Holmberg's bootstrap population has exactly the same auxiliary total as in the original population

$$X^* = \sum_{i \in s} d_i x_i = \sum_{i \in s} x_i / \pi_i = X. \tag{6}$$

Furthermore, under the same condition, we have

$$Y^* = \sum_{i \in s} d_i y_i = \hat{Y}_{HT}, \qquad \pi_{i^*} = \pi_i \tag{7}$$

and

$$E^*(\hat{Y}^*) = E^* \left(\sum_{i^* \in s^*} y_{i^*} / \pi_{i^*} \right) = \sum_{i^* \in U^*} \pi_{i^*} y_{i^*} / \pi_{i^*} = Y^* = \hat{Y}_{HT} . \tag{8}$$

However, the request $r_i = 0$ for all sampled units is obviously unrealistic in the applications, so that the properties of (6)–(8) above hold approximatively in a general case. Moreover the algorithm may result computationally heavy due to the randomisation in bootstrap population step 1. In fact n Bernoulli random variables have to be simulated in order to compute the weights d_i and to provide a bootstrap population U^*. Hence, in the general case of r_i not necessary null, the original Holmberg's πPS bootstrap actually concerns a class $\mathcal{U} = \{U_h^*, h = 1 \ldots 2^n\}$ of 2^n possible bootstrap populations. These bootstrap populations differ from each other with respect to the size N^*, to the auxiliary total X^* and to the set of inclusion probabilities π_{i^*} as shown in Example 1.

Example 1. Let U be of size $N = 10$ and the auxiliary variable x take values $x_i = i$ $(i = 1 \cdots 10)$ so that $X = 55$. Set $n = 4$ and let $s = (5, 6, 8, 10)$ be the original sample selected from U under a πPS design with inclusion probability $\pi_i = nx_i/X$ as reported in Table 1.

Table 1. Original sample inclusion probabilities and weights in step 1 of the Holmberg algorithm

$i \in s$	π_i	$1/\pi_i$	c_i	r_i
5	0.364	2.750	2	0.75
6	0.436	2.292	2	0.292
8	0.582	1.719	1	0.719
10	0.727	1.375	1	0.375

The randomisation $d_i = c_i + Bern(r_i)$ embedded in step 1 of the Holmberg algorithm, generates a set of 16 possible bootstrap populations as displayed in Table 2, along with their size, their auxiliary total and the probability of being in fact the result U^* of the bootstrap population in step 1.

Table 2. The class of 16 bootstrap populations generated by the randomisation required by the original Holmberg algorithm

h	U_h^*	d_5	d_6	d_8	d_{10}	N_h^*	X_h^*	$P\{U_h^* = U^*\}$
1	{5,5,6,6,8,10}	2	2	1	1	6	40	0.0311
2	{5,5,6,6,8,10,10}	2	2	1	2	7	50	0.0187
3	{5,5,6,6,8,8,10}	2	2	2	1	7	48	0.0795
4	{5,5,6,6,8,8,10,10}	2	2	2	2	8	58	0.0477
5	{5,5,6,6,6,8,10}	2	3	1	1	7	46	0.0128
6	{5,5,6,6,6,8,10,10}	2	3	1	2	8	56	0.0077
7	{5,5,6,6,6,8,8,10}	2	3	2	1	8	54	0.0328
8	{5,5,6,6,6,8,8,10,10}	2	3	2	2	9	64	0.0197
9	{5,5,5,6,6,8,10}	3	2	1	1	7	45	0.0934
10	{5,5,5,6,6,8,10,10}	3	2	1	2	8	55	0.0560
11	{5,5,5,6,6,8,8,10}	3	2	2	1	8	53	0.2386
12	{5,5,5,6,6,8,8,10,10}	3	2	2	2	9	63	0.1432
13	{5,5,5,6,6,6,8,10}	3	3	1	1	8	51	0.0385
14	{5,5,5,6,6,6,8,10,10}	3	3	1	2	9	61	0.0231
15	{5,5,5,6,6,6,8,8,10}	3	3	2	1	9	59	0.0983
16	{5,5,5,6,6,6,8,8,10,10}	3	3	2	2	10	69	0.0590

As a consequence, the degree of approximation of the properties of the Holmberg's algorithm as given by equations (6) to (8) would change to a possibly large extent among the bootstrap populations included in the class $\mathcal{U}$. Since the randomisation is ultimately equivalent to select a single U^* from $\mathcal{U}$, thus in the general case when the randomisation has to be actually performed, the Holberg's πPS algorithm, beside being resource consuming, may also prove to be unstable, providing variance bootstrap estimator with high variability. In the following sections we will then propose some modifications of the original Holmberg's algorithm with the twofold purpose of simplifying it computationally and of improving the efficiency of the resulting variance estimator.

4 The 0.5 πPS-bootstrap

We first focus on simplifying the Holmberg's algorithm by eliminating the randomisation performed at step 1 in the general case of r_i not necessarily null for all $i \in s$. We consider the most natural approximation of weights d_i to the nearest integer and call it the "0.5 πPS-bootstrap"

$$d_i = \begin{cases} c_i & \text{if} \quad r_i < 0.5 \\ c_i + 1 & \text{if} \quad r_i \geq 0.5. \end{cases} \tag{9}$$

Hence, a unique bootstrap population U^* is readily reached. Moreover, it is the maximum probability U^* in the class $\mathcal{U}$ since it maximises the joint probability $\prod_{i \in s} r_i^{\varepsilon_i}(1-r_i)^{1-\varepsilon_i}$ of the n independent Bernoulli trials required by the original Holberg's algorithm in Section 3. For instance, in Example 1 the 0.5-rule would give $U^* = U_{11}^*$ which has the greatest probability of appearing among the set of the possible 16. The 0.5 πPS algorithm then continues as for Holmberg's with the remaining steps: the resampling, performed with the same original sample size n and under the same πPS design, the replication, the bootstrap distribution and the variance estimation step. The entire process eventually leads to the bootstrap variance estimator $v_{0.5}$ of $V(\hat{Y}_{HT})$.

5 The x-balanced πPS-bootstrap

Two more proposals are now introduced to foster efficiency gains in the bootstrap estimation of $V(\hat{Y}_{HT})$ by a more complete use of the auxiliary information. The two proposals are slightly different versions of the same algorithm which we will call the "x-balanced πPS-bootstrap". A balancing with respect to the known population auxiliary total is indeed suggested in constructing U^*, *i.e.* under the restriction $X^* \approx X$ [21]. Let U_0^* be the basic bootstrap population formed by sampled units each replicated $c_i = \lfloor \pi_i^{-1} \rfloor$ times. Starting from U_0^* we first propose to iteratively add sampled units $i \in s$ previously sorted in a decreasing order according to $r_i = \pi_i^{-1} - c_i$. The process ends when the bootstrap population ensuring the best approximation to

X is detected in $\mathcal{U}$, *i.e.* when $|X^* - X|$ reaches its minimum in $\mathcal{U}$. Hence, step 1 in the Holmberg's algorithm is now substituted by the following 3 sub-steps:

- for $t = 0$ let $U^*_{(0)} = U^*_0$ and $s_{(0)} = s$;
- for $t \geq 1$ select unit i_t in $s_{(t-1)}$ so that $r_{i_t} \geq r_j, \quad \forall j \in s_{(t-1)}$; add unit i_t to $U^*_{(t-1)}$ thus producing $U^*_{(t)}$; subtract unit i_t from $s_{(t-1)}$ thus producing $s_{(t)}$;
- if $|X^*_{(t)} - X| < |X^*_{(t-1)} - X|$ then $U^* = U^*_{(t)}$ and repeat these last two points, or else $U^* = U^*_{(t-1)}$.

Once the x-balanced bootstrap population U^* is constructed, the algorithm then continues as for the original Holmberg's in Section 3 with the remaining steps (resampling, replication, bootstrap distribution and variance estimation step) eventually producing the bootstrap variance estimator v_{xbal1} of $V(\hat{Y}_{HT})$.

Alternatively, an x-balanced bootstrap population is also reached by using $q_i = \pi_i^{-1}/(\lfloor \pi_i^{-1} \rfloor + 1)$ instead of r_i. Weights q_i are inspired by the D'Hondt method for chair distribution in a proportional election system [6]. They guarantee that sampled units $i \in s$ with greater c_i for equal r_i are given a priority, *i.e.* units with larger frequency in U^*_0 enter $U^*_{(t)}$ first in the algorithm above. The resulting bootstrap variance estimator will be denoted by v_{xbal2}.

The x-balanced algorithm, in both versions 1 and 2, leads to a unique U^*, still included in $\mathcal{U}$, in a number of steps less than or equal to n while the original Holmberg's algorithm always needs exactly n randomisation runs. In addition, efficiency improvements are expected from using a bootstrap population closer to the actual population with respect to the auxiliary total X. Finally, notice that all the πPS bootstrap algorithms discussed so far certainly produce a zero variance estimate when the auxiliary variable is exactly proportional to the study variable. Hence, v_H, $v_{0.5}$, v_{xbal1} and v_{xbal2} share a good property with the classical unbiased estimator v_{SYG} since $y \propto x$ leads to $V(\hat{Y}_{HT}) = 0$.

A number of further approaches to πPS bootstrap have been tentatively explored with the same purpose of computational simplification and improving efficiency. All of them have been eventually abandoned on the basis of poor simulation results since when performed with sensitive computation resources, they tend to fail due to the estimator properties perspective and *vice versa*; although they prove to be interesting from a theoretical viewpoint, they did not produce encouraging simulation results.

6 Simulation study

In order to check the empirical performance of the πPS bootstrap algorithms proposed, an extensive simulation study was carried out. Eight variance estimators are concerned:

- the traditional unbiased estimator v_{SYG} defined in (2), serving as a benchmark;
- the naïve bootstrap estimator v_{naive} given by (4);
- the bootstrap estimator v_H defined by (5) offered by the original Holmberg's algorithm discussed in Section 3;

- the bootstrap estimator $v_{0.5}$ produced by the 0.5 πPS-bootstrap proposed in Section 4;
- the two variance estimators v_{xbal1} and v_{xbal2} corresponding to the two versions of the x-balanced πPS-bootstrap proposed in Section 5.

The set of variance estimators compared has been completed by adding two *approximated* πPS variance estimators. By using an approximated variance estimator for $V\left(\hat{Y}_{HT}\right)$ the computation of the joint inclusion probability π_{ij} is overcome by suitably approximating them in terms of the simple inclusion probability π_i only. Hence, any approximated variance estimator is a competitor of a bootstrap solution although limited to the estimation of $V(\hat{Y}_{HT})$. Among the many available in literature, the following two approximated estimators have been chosen since they emerged as nearly unbiased and the most stable from previous empirical studies [10, 11]:

$$v_{HR} = \frac{1}{n-1} \sum \sum_{i>j \in s} \left(1 - \pi_i - \pi_j + \frac{1}{n} \sum_{i \in U} \pi_i^2\right) \left(\frac{y_i}{\pi_i} - \frac{y_j}{\pi_j}\right)^2, \tag{10}$$

requiring the simple inclusion probability π_i known for all population units $i \in U$ [9] and

$$v_{BM} = \sum_{i \in s} \left(\frac{n}{n-1} - \pi_i\right) \left(\frac{y_i}{\pi_i} - \frac{\hat{Y}_{HT}}{n}\right)^2, \tag{11}$$

which is readily computed from sample data.

Samples were simulated under the πPS Rao-Sampford design [15,16] which guarantees the estimator v_{SYG} to be unbiased and uniformly positive. Moreover, recoursive formulae are available to compute the joint inclusion probability π_{ij} for any sample size n so that the exact value of v_{SYG} can be computed for simulation comparisons. In addition, it is already implemented in SAS and R (`pps` and `sampling` packages). On the other hand, it is a rejective design meaning that the sample selection is actually performed with replacement, but the sample is accepted only if it contains all distinct units. Otherwise, the sample is totally rejected at the occurrence of a duplicated unit and the selection must restart from the beginning. Hence, the rejection probability increases with the sampling fraction n/N and with the presence of a large π_i making the Rao-Sampford πPS design inconvenient for simulation purposes when sampling fractions are greater than 10%.

Three standard Monte Carlo (MC) performance indicators have been computed:

- the MC Relative Bias,

$$RB = \frac{E_{MC}(v) - V(\hat{Y}_{HT})}{V(\hat{Y}_{HT})}; \tag{12}$$

- the MC Relative Efficiency of a variance estimator v with respect to v_{SYG},

$$Eff = \frac{MSE_{MC}(v_{SYG})}{MSE_{MC}(v)}; \tag{13}$$

- the MC coverage of 95% confidence intervals, computed according to the percentile bootstrap method with regard to the boostrap variance estimator and according to a standard Student's t for the remaining variance estimators considered.

Three levels of sampling fraction $f = n/N$ (5%, 10% and 15%) have been explored under the restriction $\pi_i < 1$ for all population units for the sake of simplicity. The number of simulation runs (between 1000 and 10,000) has been used to control the Monte Carlo error on the basis of the unbiased estimators according to the rule:

$$\frac{\left|E_{MC}\left(\hat{Y}_{HT}\right) - Y\right|}{Y} < 1\% \quad \text{and} \quad \frac{\left|E_{MC}(v_{SYG}) - V(\hat{Y}_{HT})\right|}{V(\hat{Y}_{HT})} < 3\%. \tag{14}$$

An array of different scenarios has been explored by simulating from 5 populations. Two natural populations, named MU100 and MU100CS, have been randomly selected from the conventional Swedish dataset MU281 [17] by using different sets of study and auxiliary variables y and x. Three artificial populations, called Gamma-Normal and denoted by GN1, GN2 and GN3, have also been considered which allow for a total control of the experimental conditions. The auxiliary variable x was generated with Gamma distribution and increasing levels of variability as measured by the coefficient of variation cv_x. The study variable y was produced conditionally to x under the etheroschedastic model $y_i|x_i = ax_i + N(0, x_i)$. The correlation between x and y has been kept close to 0.9 since high correlation suggests the use of a πPS design, *i.e.* a complex sampling, as opposed to a simpler design like SRS. The simulation setup is summarised in Table 3 and displayed in Figure 1 and 2.

Table 3. Characteristics of natural and artificial populations simulated

Population	N	cv_y	cv_x	ρ_{xy}
MU100	100	1.107	1.015	0.9931
MU100CS	100	0.325	0.527	0.2829
GN1	100	0.529	0.598	0.897
GN2	100	0.981	1.122	0.916
GN3	100	1.419	1.692	0.928

Simulation results are reported in Tables 4 to 7. The evident poor performance of the naïve bootstrap confirms the need for suitable modified algorithms applied to non-*iid* situations. Simulation results also show the good performance of the modified algorithms proposed in Section 4 and 5 in terms of bias and stability as compared with all the other estimators considered, both bootstrap (numeric) and approximated (analytic). In some cases they allow for efficiency gains, greater or equal to 7%, with respect to the traditional unbiased estimator. The bootstrap estimator $v_{0.5}$ can perform noticeably better than the Holmberg estimator v_H (as for instance in populations GN3 and GN2 for the sampling fraction $f = 0.15$). It outperforms both v_{xbal1} and v_{xbal2} in 7 out of the 12 cases explored. The two x-balanced bootstrap estimators perform

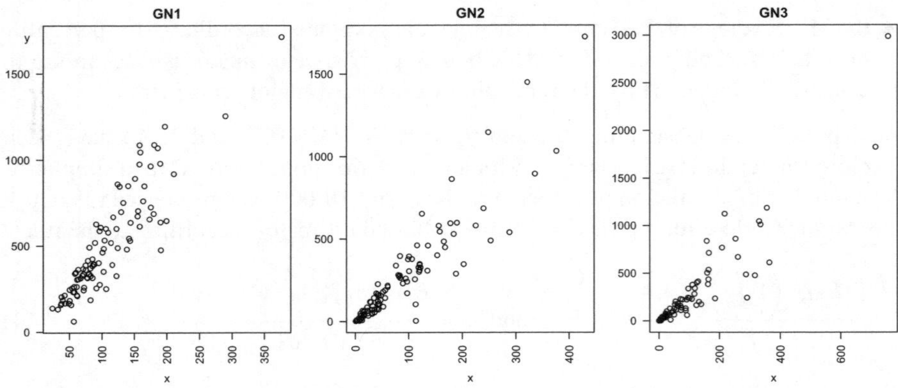

Fig. 1. Scatter plot for artificial populations

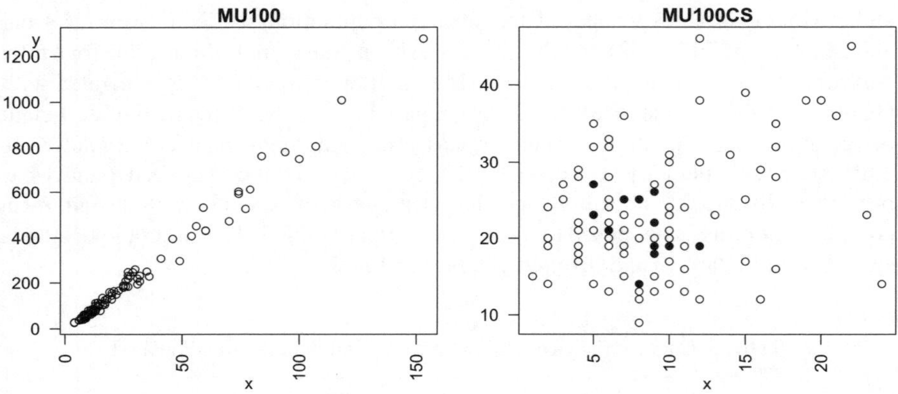

Fig. 2. Scatter plot for natural populations

quite similarly with a slight superiority of v_{xbal1} over v_{xbal2}. Neither the sampling fraction nor the variability of y and x appear to affect the pattern of the relative efficiency between the bootstrap variance estimators and the traditional unbiased estimator v_{SYG}. On the other hand, the proposed πPS-bootstrap algorithms show good empirical results even in a context inappropriate for πPS sampling, as for instance in population MU100CS with a low level of ρ_{xy} and cv_x. With regard to confidence intervals, the πPS-bootstraps proposed present a coverage of between 77% and 93% versus a 95% confidence level. As a matter of fact, in this simulation the bootstrap percentile method tends to produce confidence intervals systematically shorter than the standard student's t method. Both the x-balanced boostrap estimators reach a coverage always better than v_H. The coverage offered by the 0.5 πPS-bootstrap clearly improves for increasing sampling fractions. As a conclusion, the 0.5 πPS-bootstrap is recommended since it simplifies the original Holmberg algorithm while retaining good statistical properties.

Table 4. Simulation results for natural populations

estimator	*MU100 f* = 0.05			*MU100 f* = 0.10			*MU100CS f* = 0.05			*MU100CS f* = 0.10		
	RB	*Eff*	*coverage*	*RB*	*Eff*	*coverage*	*RB*	*Eff*	*coverage*	*RB*	*Eff*	*coverage*
SYG	−1.62	1.0000	94.1	0.85	1.0000	94.5	−2.46	1.0000	88.2	−1.18	1.0000	87.3
H	0.23	0.9671	79.3	1.76	1.0182	76.9	−2.42	1.0191	78.3	−1.24	1.0178	83.6
0.5	0.41	1.0031	79.9	5.58	1.0093	81.2	−2.20	1.0154	78.0	−1.82	1.0300	82.4
xbal1	0.53	0.9939	82.4	1.12	1.0740	89.2	−1.43	0.9842	78.1	−0.94	1.0191	83.4
xbal2	0.13	0.9872	82.3	3.00	1.0108	87.3	−2.12	0.9960	78.5	0.37	0.9995	83.5
naive	8.22	0.8747	85.1	23.46	0.6538	92.3	1.15	0.9859	79.0	7.00	0.9217	85.1
BM	0.40	0.9957	94.2	3.45	1.0287	94.7	−2.14	1.0102	88.8	−0.83	1.0143	87.7
HR	−1.48	0.9991	94.1	1.53	0.9937	94.5	−2.44	0.9998	88.3	−1.06	0.9983	87.4

Table 5. Simulation results for artificial population GN1

estimator	*GN1 f* = 0.05			*GN1 f* = 0.10			*GN1 f* = 0.15		
	RB	*Eff*	*coverage*	*RB*	*Eff*	*coverage*	*RB*	*Eff*	*coverage*
SYG	2.01	1.0000	96.0	2.27	1.0000	94.6	0.46	1.0000	94.4
H	3.21	0.9760	84.1	3.39	0.9828	89.6	1.50	0.9718	90.0
0.5	2.84	0.9960	83.6	5.00	0.9249	89.3	5.44	0.8680	90.4
xbal1	3.16	0.9784	84.7	3.59	0.9698	90.2	2.97	0.9587	91.4
xbal2	3.35	0.9785	85.0	3.66	0.9663	89.8	2.17	0.9557	91.3
naive	8.63	0.8813	85.2	16.36	0.7302	91.4	22.64	0.5556	94.0
BM	3.24	0.9834	96.2	3.66	0.9851	94.8	1.83	0.9993	94.7
HR	1.99	1.0009	96.0	2.23	1.0020	94.6	0.38	1.0030	94.4

Table 6. Simulation results for artificial population GN2

estimator	*GN2 f* = 0.05			*GN2 f* = 0.10			*GN2 f* = 0.15		
	RB	*Eff*	*coverage*	*RB*	*Eff*	*coverage*	*RB*	*Eff*	*coverage*
SYG	2.84	1.0000	95.8	0.98	1.0000	95.9	0.64	1.0000	95.4
H	4.57	0.9963	82.1	2.19	1.0055	89.7	1.87	0.9915	86.8
0.5	4.58	0.9925	82.3	5.61	0.9394	89.2	−2.44	1.1301	90.6
xbal1	4.45	0.9996	82.3	3.02	0.9914	90.9	−0.43	1.0804	91.6
xbal2	4.28	1.0011	82.3	2.49	0.9756	91.3	3.40	0.9424	91.1
naive	13.76	0.8556	83.1	24.00	0.6554	94.1	40.33	0.4302	95.6
BM	4.89	0.9997	95.8	3.07	1.0064	94.8	3.18	1.0374	95.5
HR	2.81	1.0033	95.8	0.91	1.0061	94.6	0.49	1.0125	95.4

Table 7. Simulation results for artificial population GN3

estimator	*GN3* $f = 0.05$			*GN3* $f = 0.10$		
	RB	*Eff*	*coverage*	*RB*	*Eff*	*coverage*
SYG	2.04	1.0000	95.7	−2.29	1.0000	93.8
H	4.45	1.0080	84.0	0.20	0.9012	84.4
0.5	0.78	1.0769	85.8	−1.27	1.0421	86.6
xbal1	2.50	1.0350	84.9	−0.55	1.0140	88.0
xbal2	4.34	0.9802	84.0	1.50	0.9298	88.6
naive	19.13	0.8212	86.8	40.57	0.4331	95.0
BM	5.26	1.0279	95.8	2.46	1.0090	95.6
HR	1.92	1.0184	95.7	−2.43	1.0330	94.4

7 Conclusions

In complex survey sampling the bootstrap method may be a simple numeric solution for variance estimation, which is an essential moment in the estimator's accuracy assessment, and for providing confidence intervals. On the other hand appropriate modifications of the original Efron's bootstrap ought to be considered in order to address the complexity of the sampling design. In this paper, focusing on a without replacement πPS design and on the Horvitz-Thompson estimator for the population total, some πPS-bootstrap algorithms have been discussed. Starting from the Holmberg's algorithm which is based on the appealing concept of bootstrap population and has exhibited encouraging results in previous empirical work, three modified algorithms have been proposed with the twofold purpose of computational simplification and improving efficiency. The performance of the πPS bootstraps proposed have been checked via a simulation study concerning both real and artificial data. Empirical results show that a bootstrap variance estimator is an adequate solution provided that the non-*iid* features are effectively addressed. The three modified algorithms proposed show good empirical performance in the linear case for being nearly as efficient as the customary unbiased Sen-Yeates-Grundy estimator of the variance of the Horvitz-Thompson estimator. Among the algorithms proposed, the 0.5 πPS-bootstrap is recommended since it definitely simplifies the Holmberg algorithm whilst ensuring good statistical properties. Since a πPS-bootstrap variance estimator is a general tool for applying to any estimator of any population parameter, no matter what its analytic complexity, the results presented in this paper encourage future research concerning non-linear cases and different complex designs.

References

1. Aires, N.: A guide to Fortran programs to calculate inclusion probabilities for conditional Poisson sampling and Pareto πps sampling designs. Computational Statistics **19**, 337–345 (2004)

2. Brewer, K.R.W., Hanif, M.: Sampling with Unequal Probabilities. Springer-Verlag, New York (1983)
3. Chao, M.T., Lo, A.Y.: A bootstrap method for finite population. Sankhya: The Indian Journal of Statistics **47**(A), 399–405 (1985)
4. Chao, M.T., Lo, A.Y.: Maximum likelihood summary and the bootstrap method in structured finite populations. Statistica Sinica **4**, 389–406 (1994)
5. Chaudhuri, A., Vos, J.W.E.: Unified Theory and Strategies of Survey Sampling. North-Holland, Amsterdam (1988)
6. D'Hondt, V.: La représentation proportionnelle des partis par un électeur. Gent (1878)
7. Efron, B.: Bootstrap methods: another look at the jackknife. Annals of Statistics **7**, 1–26 (1979)
8. Gross, S.: Median estimation in sample surveys. *Proceedings of Section on Survey Research Methods*. American Statistical Association, Washington, 181–184 (1980)
9. Hartley, H.O., Rao, J.N.K.: Sampling with unequal probability and without replacement. The Annals of Mathematical Statistics **32**, 350–374 (1962)
10. Haziza, D., Mecatti, F., Rao, J.N.K.: Comparisons of variance estimators under Rao-Sampford method: a simulation study. *Proceedings of the Survey Methods Section*. American Statistical Association, Toronto, 3638–3643 (2004)
11. Haziza, D., Mecatti, F., Rao, J.N.K.: Evaluation of some approximate variance estimators under the Rao-Sampford unequal probability sampling design. Metron **1**, 89–106 (2008)
12. Holmberg, A.: A bootstrap approach to probability proportional to size sampling. *Proceedings of Section on Survey Research Methods*. American Statistical Association, Washington, 378–383 (1998)
13. McCarthy, P.J., Snowden, C.B.: The bootstrap and finite population sampling. Vital and Health Statistics **2**(95), U.S. Government Printing Office, Washington (1985)
14. Rao, J.N.K., Wu, C.F.J.: Resampling inference with complex survey data. Journal of the American Statistical Association **83**(401), 231–241 (1988)
15. Rao, J.N.K.: On two simple schemes of unequal probability sampling without replacement. Journal of the Indian Statistical Association **3**, 173–180 (1965)
16. Sampford, M.R.: On sampling without replacement with unequal probabilities of selection. Biometrika **54**(3/4), 499–513 (1967)
17. Särndal, C.E., Swensson, B., Wretman, J.: Model Assisted Survey Sampling. Springer-Verlag, New York (1992)
18. Sen, A.R.: On the Estimate of the Variance in Sampling with Varying probabilities. Journal of the Indian Society of Agricultural Statistics **7**, 119–127 (1953)
19. Shao, J., Tu, D.: The jackknife and bootstrap. Springer-Verlag, New York (1995)
20. Sitter, R.R.: A resampling procedure for complex survey data. Journal of the American Statistical Association **87**(419), 755–765 (1992)
21. Tillé, Y.: Sampling algorithms. Springer-Verlag, New York (2006)
22. Yates, F., Grundy, P.M.: Selection Without Replacement from Within Strata with Probability Proportional to Size. Journal of the Royal Statistical Society, Ser. B, **15**, 235–261 (1953)

Fast Bayesian functional data analysis of basal body temperature

James M. Ciera

Abstract. In many clinical settings, it is of interest to monitor a bio-marker over time for a patient in order to identify or predict clinically important features. For example, in reproductive studies that involve basal body temperature, a low, high point or sudden changes on the trajectory have important clinical significance in determining the day of ovulation or in causing dysfunctional cycles. It is common to have patient databases with a huge quantity of data and patient information is characterised with cycles that have sparse observations. If the main interest is to make predictions, it is crucial to borrow information across cycles and among patients. In this paper, we propose the use of fast and efficient algorithms that rely on spareness-favouring hierarchical priors for P-spline basis coefficients to aid estimation of functional trajectories. Using the basal body temperature data, we present an application of the Relevant Vector Machine method that generates sparse functional linear and linear mixed models that can be used to rapidly estimate individual-specific and population average functions.

Key words: functional linear models, MAP estimates, ovulation, Relevance Vector Machine, sparsity

1 Introduction

In many clinical studies, measurements are collected repeatedly from many subjects over a period of time. Using massive datasets, physicians require fast automated tools to estimate data trajectories and predict clinically important events for a patient. For example, in reproductive studies that involve basal body temperature, a low, high point or sudden changes on the trajectory have important clinical significance in determineng the day of ovulation or identifying dysfunctional menstrual cycles [1]. Borrowing information from different subjects is crucial when observations are sparse and the interest is in prediction. Therefore, there is a need for fast algorithms for estimating functional trajectories while borrowing information from other patients concerning the shape and location of features in the function.

Functional data analysis (FDA) methods are commonly used to estimate curves but rely on a large number of basis functions [17]. Early work in functional data analysis used different smoothing methods to estimate individual and population mean trajec-

Mantovan, P., Secchi, P. (Eds.): Complex Data Modeling and Computationally Intensive Statistical Methods

tories. For example, [26] used semiparametric stochastic models while accounting for the within-subject correlation using a stationary or non-stationary Gaussian process. Brumback and Rice used non-parametric methods to estimate both population average and subject-specific profiles such that the smoothing splines were represented using a mixed model [2]. Rice and Wu used a low-rank spline basis approach to model individual curves as spline functions with random coefficients [18]. Guo linked the smoothing splines and mixed models to estimate subject-specific curves [11]. Recently, Durban, Thompson and Crainiceanu introduced non-Bayesian and Bayesian approaches that rely on a trade-off between spline regression and smoothing splines [7, 8, 22]. However, most of these approaches are computationally intensive and it is common to encounter computational problems when the dimension of the basis functions and the number of subjects becomes large.

Various model reduction methods have been introduced for functional linear models. For example, in Bayesian statistics, selection of the basis functions commonly relies on computationally intensive reversible jump algorithms [10] or stochastic search variable selection methods [21]. In a functional linear mixed model, selection of the random effects has complications since the null hypothesis lies on the boundary of the parameter space and the classical likelihood ratio test statistic is no longer valid. To solve this problem [16] and [14] introduced alternative approaches to reduce the dimension of the random effects model. Recent methods use functional principal component analysis [7, 12, 25]. The approaches have good performance in modest dimensional models with moderate numbers of subjects, but rapidly becomes computationally unfeasible as the number of subjects and candidate predictors increases.

In Bayesian statisics, these problems raise a practical motivation for fast approximate Bayes methods that can bypass MCMC while maintaining some of the benefits of a Bayesian analysis. The Relevant Vector Machine (RVM) [23] is one of these fast Bayesian methods that promotes sparseness in estimation of basis coefficients, providing a more flexible alternative to Support Vector Machines (SVM) [3]. RVM is based on empirical Bayes methodology and penalises the basis coefficients through a scale mixture of normals prior, which is carefully-chosen so that maximum a posteriori (MAP) estimates of many of the coefficients are zero. This provides a natural mechanism in selection of the basis functions leading to a sparse model that is fast to compute. The RVM approach can be used to generate a sparse linear model [4, 13] as well as a linear mixed model [5]. Using the basal body temperature data [6], we present an application of RVM methods to generates sparse functional linear and linear mixed models that can be used to rapidly estimate individual-specific and population average functions.

The subsequent sections are as follows; Section 2 discusses the RVM procedure, while Section 3 contains results based on the bbt data from the European fecundability study.

2 Methods

Let $y_i = (y_{i1}, \ldots, y_{iT_i})'$ and $z_i = (z_{i1}, \ldots, z_{iT_i})'$ be vectors for the response and covariates for the ith woman. A functional model is represented as

$$y_{it} = f_i(z_{it}) + \epsilon_{it}, \quad \epsilon_{it} \sim N(0, \sigma_\epsilon^2), \quad t = 1, \ldots, T_i, \quad i = 1, \ldots, N, \tag{1}$$

where $f_i(.)$ is a smooth function at z_{it} for subject i, and ϵ_{it} is a measurement error. The functional model in (1) can be represented as a linear or linear mixed model depending on whether the interest is on either subject-specific curves or both the population and subject-specific curves.

2.1 RVM in linear models

When the interest is to model the subject-specific curves, the functional model in (1) can be represented as a linear model such that the smoothing function is described as a linear combination of M basis functions

$$f_i(z_{it}) = \sum_{j=1}^{M} \beta_{ij}\varphi_j(z_{it}) = \mathbf{x}_{it}'\boldsymbol{\beta}_i, \tag{2}$$

where $\boldsymbol{x}_{it} = (x_{it1}, \cdots, x_{itM})'$ are the values of the basis functions at z_{it}, parameter β_{ij} is the coefficient for the jth basis function $\varphi_j(.)$ and $\boldsymbol{\beta}_i = (\beta_{i1}, \cdots, \beta_{iM})'$. The basis functions φ_j can be generated using numerous methods that have been discussed in the literature (e.g. [9, 19]).

The priors are $\beta_{ij} \sim N(0, \alpha_j^{-1})$, $\sigma_\epsilon^{-2} \sim Gamma(a, b)$ and $\alpha_j \sim Gamma(c, d)$. The parameters $\boldsymbol{\alpha} = (\alpha_1, \cdots, \alpha_M)'$ and σ_ϵ^{-2} are computed from the data as *maximum a posteriori* (MAP) estimates. Since these MAP estimates are estimated and shared among all the subjects, this leads to borrowing of strength across subjects in estimating subject-specific functions [4, 13]. To promote sparseness over the model coefficients $\boldsymbol{\beta}_i$, the hyperparameters c and d are set close to zero leading to a distribution with a large spike concentrated at zero and a heavy right tail. The basis functions for which α_j is in the right tail have coefficients that are strongly shrunk toward zero.

Inference in Bayesian data analysis is based on the posterior distribution of the parameters. In the RVM approach, the posterior density is based on the conditional distribution

$$p(\boldsymbol{\beta}, \boldsymbol{\alpha}, \sigma_\epsilon^{-2}|\boldsymbol{Y}) = p(\boldsymbol{\alpha}, \sigma_\epsilon^{-2}|\boldsymbol{Y}) \prod_{i=1}^{N} p(\boldsymbol{\beta}_i|\boldsymbol{y}_i, \boldsymbol{\alpha}, \sigma_\epsilon^{-2}),$$

where $\boldsymbol{\beta} = (\boldsymbol{\beta}_1, \cdots \boldsymbol{\beta}_N)$. The posterior density $p(\boldsymbol{\beta}_i|\boldsymbol{y}_i, \boldsymbol{\alpha}, \sigma_\epsilon^{-2})$ is a multivariate normal

$$p(\boldsymbol{\beta}_i|\boldsymbol{Y}, \boldsymbol{\alpha}, \sigma_\epsilon^{-2}) = N(\boldsymbol{\beta}_i; \hat{\boldsymbol{\nu}}_i, \hat{\boldsymbol{\Sigma}}_i), \tag{3}$$

where $\hat{\boldsymbol{\mu}}_i = \sigma_\epsilon^{-2}\hat{\boldsymbol{\Sigma}}_i\boldsymbol{X}_i'\boldsymbol{y}_i$ and $\hat{\boldsymbol{\Sigma}}_i = (\boldsymbol{A} + \sigma_\epsilon^{-2}\boldsymbol{X}_i'\boldsymbol{X}_i)^{-1}$ such that $\boldsymbol{A} = diag\{\alpha_1, \cdots, \alpha_M\}$ and $\boldsymbol{X}_i = (\boldsymbol{x}_{i1}, \cdots, \boldsymbol{x}_{iM})'$.

The density for $p(\boldsymbol{\alpha}, \sigma_\epsilon^{-2}|\boldsymbol{Y})$ is difficult to express analytically. To compute the estimates for $\boldsymbol{\alpha}$ and σ_ϵ^{-2}, we assume that the modes for $p(\boldsymbol{\alpha}, \sigma_\epsilon^{-2}|\boldsymbol{Y})$ and $p(\boldsymbol{Y}|\boldsymbol{\alpha}, \sigma_\epsilon^{-2})$ are equivalent and hence the MAP estimates for $p(\boldsymbol{\alpha}, \sigma_\epsilon^{-2}|\boldsymbol{Y})$ are equivalent to the MLE estimates from $p(\boldsymbol{Y}|\boldsymbol{\alpha}, \sigma_\epsilon^{-2})$ [13]. The estimates for $\boldsymbol{\alpha}$ and σ_ϵ^{-2} are computed from the marginal likelihood $p(\boldsymbol{Y}|\boldsymbol{\alpha}, \sigma_\epsilon^{-2})$, obtained after integrating out $\boldsymbol{\beta}_i$ from $p(\boldsymbol{Y}|\boldsymbol{\beta}_i, \sigma_\epsilon^{-2})$ such that

$$p(\boldsymbol{Y}|\boldsymbol{\alpha}, \sigma_\epsilon^{-2}) = \int \prod_{i=1}^{N} p(\boldsymbol{Y}|\boldsymbol{\beta}_i, \sigma_\epsilon^{-2}) p(\boldsymbol{\beta}_i|\boldsymbol{\alpha}) d\boldsymbol{\beta}_i.$$

This results to a multivariate normal density $N(\boldsymbol{y}_i; \boldsymbol{0}, \boldsymbol{C}_i)$ where $\boldsymbol{C}_i = \sigma_\epsilon^2 \boldsymbol{I}_{T_i} + \sum_{j=1}^{M} \alpha_j^{-1} \boldsymbol{x}_{ij} \boldsymbol{x}_{ij}^{'}$. The estimate for σ_ϵ^{-2} is

$$\hat{\sigma}_\epsilon^{-2} = \frac{\sum_{i=1}^{N} ||\boldsymbol{y}_i - \boldsymbol{X}_i \boldsymbol{\mu}_i||^2}{\sum_{i=1}^{N} (T_i - M - \sum_{j=1}^{M} \alpha_j \Sigma_{i,jj})}. \tag{4}$$

In many functional data analysis cases, the dimension of the design matrix $\boldsymbol{X}_i$ is large. Computation of $\hat{\boldsymbol{\mu}}_i$ requires inverting an $M \times M$ matrix while estimating $\boldsymbol{\Sigma}_i$ in equation (3). The computation process becomes slow and inefficient prompting the need for a fast and efficient method. A fast approach follows a sequential process to estimate the elements of $\boldsymbol{\alpha}$. This is based on the dependence of the log-likelihood function $\ell(\boldsymbol{\alpha}, \sigma_\epsilon^{-2}) = \log p(\boldsymbol{Y}|\boldsymbol{\alpha}, \sigma_\epsilon^{-2})$ upon the kth element of $\boldsymbol{\alpha}$ leading to the decomposing of $\ell(\boldsymbol{\alpha}, \sigma_\epsilon^{-2})$ into two parts, one with and one without the kth element of $\boldsymbol{\alpha}$.

The solutions from the resulting score equations are unfeasible to express analytically except for a simple case where $\alpha_k = \infty$ [4, 13]. To avoid complexities, we assume that $\alpha_k \ll s_{ik}$ where $s_{ik} = \boldsymbol{x}_{ik}^{'} \boldsymbol{C}_{i,-k}^{-1} \boldsymbol{x}_{ik}$, leading to an approximate estimate

$$\hat{\alpha}_k \cong \begin{cases} \frac{N}{\sum_{i=1}^{N} (q_{ik}^2 - s_{ik})/s_{ik}^2} & \text{if } \sum_{i=1}^{N} \frac{(q_{ik}^2 - s_{ik})}{s_{ik}^2} > 0, \\ \infty & \text{otherwise.} \end{cases} \tag{5}$$

where $q_{ik} = \boldsymbol{x}_{ik}^{'} \boldsymbol{C}_{i,-k}^{-1} \boldsymbol{y}_i$ and $\boldsymbol{C}_{i,-k}$ is the component of $\boldsymbol{C}_i$ without the kth basis function. Selection of the basis function involves; addition, deletion and updating $\hat{\mu}_{ik}$. Addition occurs when $\sum_{i=1}^{N} \frac{(q_{ik}^2 - s_{ik})}{s_{ik}^2} > 0$ and $\boldsymbol{X}_{ik}$ is not in the model, while an update occurs when $\boldsymbol{X}_{ik}$ is already in the model and $\sum_{i=1}^{N} \frac{(q_{ik}^2 - s_{ik})}{s_{ik}^2} > 0$. We delete $\boldsymbol{X}_{ik}$ from the model when $\sum_{i=1}^{N} \frac{(q_{ik}^2 - s_{ik})}{s_{ik}^2} < 0$. The estimating process involves computation of $\hat{\sigma}_\epsilon^{-2}$ and $\hat{\alpha}_k$ in equations (4-5) that are used to update the mean vector and covariance matrix in equation (3). For a concrete justification of this type of approximation refer to [4, 13].

2.2 Extension to linear mixed model

The discussion in the previous section only allows estimation of subject-specific curves but not the population average. To generate the population average curve we extend model (2) into a linear mixed model that can capture both the subject-specific and population average components. The functional model in equation (1) can be generalised into a functional mixed model. The smoothing function is expressed as $f_i(z_{it}) = \sum_{j=1}^{M} \beta_j \varphi_j(z_{it}) + \sum_{j=1}^{M^*} b_{ij}\phi_j(z_{it}) = \boldsymbol{x}_{it}'\boldsymbol{\beta} + \boldsymbol{w}_{it}'\boldsymbol{b}_i$, such that $\boldsymbol{\beta} = (\beta_1, \cdots, \beta_M)'$ and $\boldsymbol{b}_i = (b_{i1}, \cdots, b_{iM^*})'$ are fixed and random effects respectively. Functions $\boldsymbol{\varphi} = \{\varphi_j\}_{j=1}^{M}$ and $\boldsymbol{\phi} = \{\phi_j\}_{j=1}^{M^*}$ are basis functions generated using methods discussed in the literature [17]. This results to the classical linear mixed model,

$$\boldsymbol{y}_i = \boldsymbol{X}_i\boldsymbol{\beta} + \boldsymbol{W}_i\boldsymbol{b}_i + \boldsymbol{\epsilon}\epsilon_i, \quad \boldsymbol{b}_i \sim N(\boldsymbol{0}, \boldsymbol{\Omega}), \quad \boldsymbol{\epsilon}_i \sim N(\boldsymbol{0}, \sigma_\epsilon^2 \boldsymbol{I}_{T_i}), \quad i = 1, \ldots, N, \tag{6}$$

where $\boldsymbol{y}_i = (y_{i1}, \cdots, y_{iM})'$, $\boldsymbol{X}_i = (\boldsymbol{x}_{i1}, \ldots, \boldsymbol{x}_{iT_i})'$, $\boldsymbol{W}_i = (\boldsymbol{w}_{i1}, \ldots, \boldsymbol{w}_{iT_i})'$ and $\boldsymbol{\epsilon}_i$ is a $T_i \times 1$ vector of error terms.

Implementation of the RVM procedure requires all random components $\boldsymbol{b}_i = (b_{i1}, \cdots, b_{iM^*})'$ and $\boldsymbol{\epsilon}_i$ to be independent leading to a covariance matrix $\boldsymbol{\Omega} = diag\{\omega_1, \cdots, \omega_{M^*}\}$. The priors for the parameters are $\beta_j|\alpha_j \sim N(0, \alpha_j^{-1})$, $\alpha_j|c_1, d_1 \sim Gamma(c_1, d_1)$, $\omega_j|c_2, d_2 \sim Gamma(c_2, d_2)$ and $\sigma_\epsilon^{-2}|a, b \sim Gamma(a, b)$. Then the joint posterior distribution is approximated as a full conditional density

$$p(\boldsymbol{\Theta}|Y) = p(\boldsymbol{b}|\boldsymbol{Y}, \boldsymbol{\beta}, \boldsymbol{\omega}, \sigma_\epsilon^{-2})p(\boldsymbol{\beta}|\boldsymbol{Y}, \boldsymbol{\alpha}, \boldsymbol{\omega}, \sigma_\epsilon^{-2})p(\boldsymbol{\alpha}, \boldsymbol{\omega}, \sigma_\epsilon^{-2}|\boldsymbol{Y}), \tag{7}$$

where $\boldsymbol{\Theta} = \{\boldsymbol{\beta}, \boldsymbol{b}, \boldsymbol{\alpha}, \boldsymbol{\omega}, \sigma_\epsilon^{-2}\}$ and $\boldsymbol{b} = (\boldsymbol{b}_1, \cdots, \boldsymbol{b}_N)'$. The first two components can be expressed as multivariate normal densities but $p(\boldsymbol{\alpha}, \boldsymbol{\omega}, \sigma_\epsilon^{-2}|\boldsymbol{Y})$ cannot be expressed analytically. The posterior distribution for $\boldsymbol{\beta}$ is,

$$p(\boldsymbol{\beta}|\boldsymbol{Y}, \boldsymbol{\beta}, \boldsymbol{\omega}, \sigma_\epsilon^{-2}) = N(\boldsymbol{\beta}; \hat{\boldsymbol{\mu}}, \hat{\boldsymbol{\Sigma}}), \tag{8}$$

where $\hat{\boldsymbol{\mu}} = \hat{\boldsymbol{\Sigma}}(\sum_{i=1}^{N} \boldsymbol{X}_i'\boldsymbol{V}_i^{-1}\boldsymbol{y}_i)$ and $\hat{\boldsymbol{\Sigma}} = (\boldsymbol{A} + \sum_{i=1}^{N} \boldsymbol{X}_i'\boldsymbol{V}_i^{-1}\boldsymbol{X}_i)^{-1}$. Matrix $\boldsymbol{V}_i = \sigma_\epsilon^2 \boldsymbol{I}_{T_i} + \boldsymbol{W}_i\boldsymbol{\Omega}^{-1}\boldsymbol{W}_i'$, where $\boldsymbol{A} = diag\{\alpha_1, \cdots, \alpha_M\}$. The posterior for the random effects $\boldsymbol{b}$ is,

$$p(\boldsymbol{b}|\boldsymbol{Y}, \boldsymbol{\beta}, \boldsymbol{\omega}, \sigma_\epsilon^{-2}) = \prod_{i=1}^{N} N(\boldsymbol{b}_i; \hat{\boldsymbol{v}}_i, \hat{\boldsymbol{\Omega}}_i), \tag{9}$$

where $\hat{\boldsymbol{v}}_i = \sigma_\epsilon^{-2}\hat{\boldsymbol{\Omega}}_i\boldsymbol{W}_i'(\boldsymbol{y}_i - \boldsymbol{X}_i\boldsymbol{\Omega})$ and $\hat{\boldsymbol{\Omega}}_i = (\boldsymbol{\Omega} + \sigma_\epsilon^{-2}\boldsymbol{W}_i'\boldsymbol{W}_i)^{-1}$.

To allow fast estimation of the posterior estimates we face two problems. Firstly, when the dimensions of both $\boldsymbol{X}_i$ and $\boldsymbol{W}_i$ are large, we encounter computation problems when estimating $\hat{\boldsymbol{\mu}}$ and $\hat{\boldsymbol{v}}_i$ while inverting the covariance matrices $\hat{\boldsymbol{\Sigma}}$ and $\hat{\boldsymbol{\Omega}}_i$ in equations (8) and (9) respectively. Secondly, the posterior $p(\boldsymbol{\alpha}, \boldsymbol{\omega}, \sigma_\epsilon^{-2}|\boldsymbol{Y})$ for the variance components lacks a simple form. Potentially this can be solved by using an empirical Bayes procedure that leads to a MAP estimation approach. This results in a fast algorithm that bypasses the inversion step leading to a reduced model with dimension $m \times m$ and $m^* \times m^*$ for both $\hat{\boldsymbol{\Sigma}}$ and $\hat{\boldsymbol{\Omega}}_i$ respectively where $m \ll M$ and $m^* \ll M^*$.

The variance components $\boldsymbol{\alpha}$, $\boldsymbol{\omega}$ and σ_ϵ^{-2} can be estimated using conditional maximisation and an empirical Bayes approach parallel to [5, 13]. We use an empirical Bayes approach to obtain plug-in estimates for $\boldsymbol{\alpha}$ and $\boldsymbol{\omega}$ that favour sparseness, and many elements of $\boldsymbol{\alpha}$ and $\boldsymbol{\omega}$ are set very close to zero. Let $p(\boldsymbol{\alpha}, \boldsymbol{\omega}, \sigma_\epsilon^{-2}|\boldsymbol{Y}) \propto p(\boldsymbol{\alpha})p(\boldsymbol{\omega})p(\sigma_\epsilon^{-2})p(\boldsymbol{Y}|\boldsymbol{\alpha}, \boldsymbol{\omega}, \sigma_\epsilon^{-2})$ where the density functions $p(\boldsymbol{\alpha})$, $p(\boldsymbol{\omega})$ and $p(\sigma_\epsilon^{-2})$ are Gamma density functions. We assume that the mode for $p(\boldsymbol{\alpha}, \boldsymbol{\omega}, \sigma_\epsilon^{-2}|\boldsymbol{Y})$ is equivalent to the MLE estimates for parameters $\boldsymbol{\alpha}$, $\boldsymbol{\omega}$ and σ_ϵ^{-2} in the likelihood function $p(\boldsymbol{Y}|\boldsymbol{\alpha}, \boldsymbol{\omega}, \sigma_\epsilon^{-2}) = N(\boldsymbol{Y}; \boldsymbol{0}, \boldsymbol{C})$, where $\boldsymbol{C} = \boldsymbol{V} + \boldsymbol{XAX}'$, $\boldsymbol{X} = (\boldsymbol{X}_1, \ldots, \boldsymbol{X}_N)'$ is the design matrix and $\boldsymbol{V} = diag(\boldsymbol{V}_1, \ldots, \boldsymbol{V}_N)$. Maximising the log-likelihood function $l(\boldsymbol{\alpha}, \boldsymbol{\omega}, \sigma_\epsilon^{-2}|\boldsymbol{Y}) = \log p(\boldsymbol{Y}|\boldsymbol{\alpha}, \boldsymbol{\omega}, \sigma_\epsilon^{-2})$ is difficult. Hence, we maximise two conditional log-likelihood functions $\ell(\boldsymbol{\alpha}; \boldsymbol{\omega}, \sigma_\epsilon^{-2}) = \frac{-1}{2}\{N\log(2\pi) + \log|\boldsymbol{C}| + \boldsymbol{Y}'\boldsymbol{C}^{-1}\boldsymbol{Y}\}$ and $\ell(\boldsymbol{\omega}, \sigma_\epsilon^{-2}; \boldsymbol{\alpha}) = -1/2\sum_{i=1}^N T_i \log(2\pi) + \log|\boldsymbol{V}_i^{-1}| + (\boldsymbol{y}_i - \boldsymbol{X}_i\hat{\boldsymbol{\mu}})'\boldsymbol{V}_i^{-1}(\boldsymbol{y}_i - \boldsymbol{X}_i\hat{\boldsymbol{\mu}})$.

We maximise the conditional log-likelihood $\ell(\boldsymbol{\alpha}; \boldsymbol{\omega}, \sigma_\epsilon^{-2})$ to obtain the estimates for $\boldsymbol{\alpha}$ while the conditional log-likelihood $\ell(\boldsymbol{\omega}, \sigma_\epsilon^{-2}; \boldsymbol{\alpha})$ is maximised to obtain the estimates for $\boldsymbol{\omega}$ and σ_ϵ^{-2}. In particular, to estimate the kth element of $\boldsymbol{\alpha}$ the conditional log-likelihood function $\ell(\boldsymbol{\alpha}; \boldsymbol{\omega}, \sigma_\epsilon^{-2})$ is decomposed into two parts with and without the kth element ($\boldsymbol{\alpha}_{-k}$). The partitioned log-likelihood function becomes

$$\ell(\boldsymbol{\alpha}; \boldsymbol{\omega}, \sigma_\epsilon^{-2}) = \ell(\boldsymbol{\alpha}_{-k}; \boldsymbol{\omega}, \sigma_\epsilon^{-2}) + \frac{1}{2}\left(\log\alpha_k - \log|\alpha_k + s_k| + \frac{q_k^2}{\alpha_k + s_k}\right),$$

where $s_k = \boldsymbol{X}'_{.k}\boldsymbol{C}_{-k}^{-1}\boldsymbol{X}_{.k}$, $q_k = \boldsymbol{X}'_k\boldsymbol{C}_{-k}^{-1}\boldsymbol{Y}$. Matrix $\boldsymbol{C}_{-k}$ does not have the contribution of the kth component of $\boldsymbol{\alpha}$ such that $\boldsymbol{C}_{-k} = \boldsymbol{V} + \sum_{j\neq k}^M \alpha_j^{-1}\boldsymbol{X}_{.j}\boldsymbol{X}'_{.j}$ where $\boldsymbol{X}_{.j}$ is the jth column of matrix $\boldsymbol{X}$. The estimate for α_k is

$$\hat{\alpha}_k = \begin{cases} \frac{s_k^2}{q_k^2 - s_k} & if\ q_k^2 > s_k, \\ \infty & otherwise. \end{cases} \tag{10}$$

Depending on the values of q_k and s_k, three operations can take place on $\boldsymbol{X}_{.k}$, including addition, deletion or update of the coefficient. At the beginning of the computation process all values of $\hat{\alpha}_k = \infty$ which corresponds to an empty model. Subsequent iterations involve computation of the values of q_k and s_k and selection of a candidate $\boldsymbol{X}_{.k}$ that has the largest contribution to the log-likelihood $\ell(\boldsymbol{\alpha}; \boldsymbol{\omega}, \sigma_\epsilon^{-2})$. When $q_k^2 > s_k$ we add $\boldsymbol{X}_{.k}$ into the model, $\hat{\alpha}_k$ is updated if $\boldsymbol{X}_{.k}$ is already in the model and deletion occurs when $\boldsymbol{X}_{.k}$ is already in the model and $q_k^2 < s_k$.

Similarly, to estimate the kth element of $\boldsymbol{\omega}$ we partition $\ell(\boldsymbol{\omega}, \sigma_\epsilon^{-2}; \boldsymbol{\alpha})$ into two parts, one with and one without the kth element ($\boldsymbol{\omega}_{-k}$). The decomposed log-likelihood becomes

$$\ell(\boldsymbol{\omega}; \boldsymbol{\alpha}, \sigma_\epsilon^2) = \ell(\boldsymbol{\omega}_{-k}; \boldsymbol{\alpha}, \sigma_\epsilon^2) - \frac{1}{2}\sum_{i=1}^N\left(\log\omega_k - \log|\omega_k + s_{ik}^*| + \frac{q_{ik}^{*2}}{\omega_k + s_{ik}^*}\right)$$

where components $s_{ik}^* = \boldsymbol{w}'_{ik}\boldsymbol{V}_{i,-k}^{-1}\boldsymbol{w}_{ik}$ and $q_{ik}^* = \boldsymbol{w}'_{ik}\boldsymbol{V}_{i,-k}^{-1}(\boldsymbol{y}_i - \boldsymbol{X}_i\hat{\boldsymbol{\mu}})$ such that $\boldsymbol{V}_{i,-k} = \sigma_\epsilon^2\boldsymbol{I} + \sum_{j\neq k}^{M^*}\alpha_j^{-1}\boldsymbol{w}_{ij}\boldsymbol{w}'_{ij}$. The solutions for the resulting score functions are unfeasible to express analytically except when $\omega_k = \infty$ [13]. To avoid computation complexities we assume that $\omega_k \ll s^*_{ik}$ leading to the approximate estimate

$$\hat{\omega}_k \cong \begin{cases} \frac{N}{\sum_{i=1}^{N}(q^{*2}_{ik}-s^*_{ik})/s^{*2}_{ik}} & \textit{if } \sum_{i=1}^{N}\frac{(q^{*2}_{ik}-s^*_{ik})}{s^{*2}_{ik}} > 0, \\ \infty & \textit{otherwise.} \end{cases} \tag{11}$$

The computation of the random effects is done sequentially. This involves three operations: add, update and delete. Addition occurs when $\sum_{i=1}^{N}\frac{(q^{*2}_{ik}-s^*_{ik})}{s^{*2}_{ik}} > 0$ and $\boldsymbol{w}_{ik}$ is not in the model, while an update occurs when $\boldsymbol{w}_{ik}$ is in the model and $\sum_{i=1}^{N}\frac{(q^{*2}_{ik}-s^*_{ik})}{s^{*2}_{ik}} > 0$. Deletion occurs when $\sum_{i=1}^{N}\frac{(q^{*2}_{ik}-s^*_{ik})}{s^{*2}_{ik}} < 0$ and $\boldsymbol{w}_{ik}$ is currently in the model. Updates for $\boldsymbol{s}^*$ and $\boldsymbol{q}^*$ are based on the values from the previous iteration. The final model has most $\omega_j = \infty$ corresponding to $b_{ij} = 0$ for all i. For more details refer to [5, 13]. The MLE estimate for σ^2_ϵ is

$$\hat{\sigma}^2_\epsilon = \frac{\sum_{i=1}^{N}||\boldsymbol{y}_i - \boldsymbol{X}_i\hat{\boldsymbol{\mu}} - \boldsymbol{W}_i\hat{\boldsymbol{v}}_i||^2}{\sum_{i=1}^{N}(T_i - M + \sum_{j=1}^{M}\omega_j\hat{\Omega}_{ijj})} \quad i = 1, \ldots N \text{ and } j = 1, \ldots M. \tag{12}$$

The sequential computation of both $\boldsymbol{\alpha}$ and $\boldsymbol{\omega}$ allows computation problems encountered to be overcome while inverting matrices $\hat{\boldsymbol{\Sigma}}$ and $\hat{\boldsymbol{\Omega}}_i$. However, the process requires a lot of iterations to reach convergence. Hence, to accelerate convergence we choose and update $\boldsymbol{\alpha}$ and $\boldsymbol{\omega}$ that lead to the largest increase in the log-likelihood functions $\ell(\boldsymbol{\alpha}; \boldsymbol{\omega}, \sigma^{-2}_\epsilon)$ and $\ell(\boldsymbol{\omega}; \boldsymbol{\alpha}, \sigma^2_\epsilon)$ respectively. The algorithm is computationally demanding when working with large datasets but is faster and more efficient relative to an approach that requires inversion of $M \times M$ and $M^* \times M^*$ matrices of $\hat{\boldsymbol{\Sigma}}$ and $\hat{\boldsymbol{\Omega}}_i$ respectively. The final model has m fixed effects and m^* random effects such that $m \ll M$ and $m^* \ll M^*$.

3 Results: application to bbt data

The research is motivated by the basal body temperature data from the European fecundability study [6]. There were 880 women in the study aged between 18 and 40 years, who were not taking hormonal medications or drugs affecting fertility, and had no known impairment of fecundity. The subjects kept daily records of cervical mucus or basal body temperature measurements from at least one menstrual cycle, and they recorded the days during which intercourse and menstrual bleeding occurred. For more details about the study, refer to [6]. In this paper we considered bbt measurements from cycles that had an identified ovulation day based on the three over six rule [15]. In total, we consider data for 520 menstrual cycles where each subject contributes data for one menstrual cycle.

Typically a standard bbt curve has a biphasic shape that is characterised with three phases representing the pre-ovulation, ovulation and post-ovulation periods. Identification of the ovulation day was based on the three over six rule [15] or a dip that is followed by a sharp rise in bbt [6]. Practically it is common to observe many menstrual cycles with wide fluctuations in bbt measurements resulting from false nadirs and peaks. Hence, it is difficult to replicate a standard bbt pattern from data collected from many cycles. Hormonal fluctuation is one among many causes that can interfere with the pattern of a bbt curve. Other causes include reduce sleep, sleep disturbances, ambient bedroom temperature, food ingestion and fluctuating emotional state [6].

Other functional data analysis methods have been proposed to model the bbt data. For example, [20] proposed a Bayesian semiparametric model based on nonparametric contamination of a linear mixed effects model. The implementation of this approach relies on a highly computational intensive MCMC algorithm and our goal is to obtain a fast approximate Bayes approach that can be implemented much more rapidly, while obtaining smooth bbt trajectories. Hence, to compare the performance of the RVM approach with an MCMC based approach, we used the subject-specific approach [24] instead of the method used in [20] since it cannot generate smooth curves.

3.1 Subject-specific profiles

To implement the RVM procedure for a linear model, we used the cubic B-splines [17] to generate the basis functions $\boldsymbol{\varphi}$. Following the Wand and Ormerod [24] approach, the basis functions were generated using the standardised values of time covariate (z_i). The number of the generated basis functions was 27 cubic B-splines with 23 interior knots. In addition, we added two columns containing $1's$ and z_i (i.e. $\{1, z_{it}\}_{t=1}^{T_i}$). Hence, the dimension for the design matrix $\boldsymbol{X}_i$ is $T_i \times M$ where $M = 29$.

Table 1 presents parameter estimates for both RVM and MCMC based procedures for two cycles. The MCMC based curves are estimated using 29 non-zero basis coefficients while the RVM method uses only three non-zero basis coefficients. Concerning time factor, the MCMC based method takes 19.30 and 16.34 seconds while the RVM

Table 1. Parameter estimates for two bbt cycles

	Parameter estimates			
Basis no.	RVM_1	RVM_2	$MCMC_1$	$MCMC_2$
1	0.0000	0.0000	−0.0361	0.0251
2	0.6837	0.6604	0.6473	0.6313
3	0.0000	0.0000	−0.0821	0.0315
4	0.0000	0.0000	−0.1751	−0.3117
.	.	.	.	.
.	.	.	.	.
.	.	.	.	.
25	0.0000	0.0000	4.1038	0.4511
26	0.0000	0.0000	−0.6272	1.4286
27	9.2149	6.1882	7.8599	3.7075
28	2.2945	0.0000	2.1504	1.9697
29	0.0000	1.3562	−0.2917	1.2575
Time spent	0.59 sec	0.57 sec	19.30 sec	16.34 sec

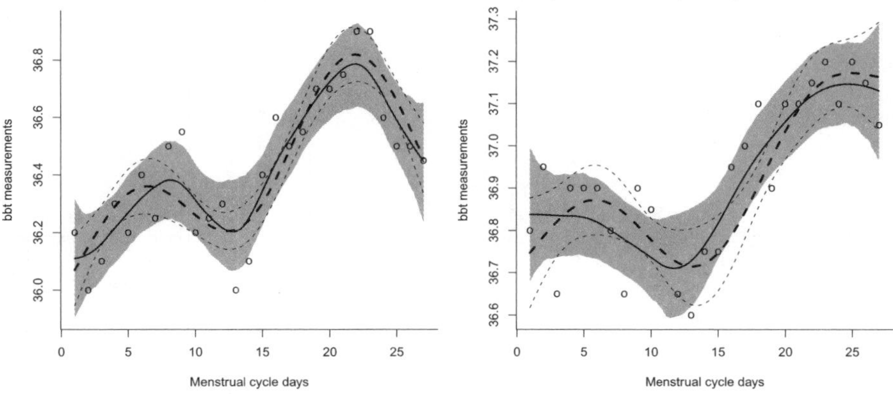

Fig. 1. Plots of bbt curves

method takes 0.59 and 0.57 seconds to estimate the two bbt curves respectively. Figure 1 presents estimated curves for the two bbt cycles using the two procedures. The continuous and dotted lines represent the estimated curves generated by the MCMC based and RVM methods respectively. The two sets of thin dotted lines and the grey region in the plots represent the credible band for the RVM and the MCMC based methods respectively. Hence, the RVM method takes less time relative to the MCMC based method given that the quality of the fitted curves is almost the same.

3.2 Subject-specific and population average profiles

Similarly, to implement the RVM procedure for a linear mixed model, the cubic B-splines [17] were used to generate the basis functions $\boldsymbol{\varphi}$ and $\boldsymbol{\phi}$. Following the Wand and Ormerod approach [24], the two sets of the basis functions were generated based on the standardised values of time covariates (z_i). Both design matrices $\boldsymbol{X}_i$ and $\boldsymbol{W}_i$, have a total of $M = M^* = 29$ columns. The first two columns in both matrices contain $1's$ and z_i (i.e. $\{1, z_{it}\}_{t=1}^{T_i}$) while the remaining columns are generated from 27 cubic B-splines with 23 interior knots. Hence, both $\boldsymbol{X}_i$ and $\boldsymbol{W}_i$ matrices have dimensions $T_i \times 29$ such that $M = M^* = 29$.

The RVM procedure for the linear mixed model and the MCMC based procedures were implemented on data for the 520 bbt cycles. The final RVM model has $m = 3$ fixed effects (basis functions: 1, 2 and 28) and $m^* = 10$ random effects (basis functions: 1, 2, 28, 29, 9, 14, 27, 12, 11 and 16). Figure 2 shows some estimated bbt curves from six randomly selected subjects based on the RVM method. The continuous line represents the estimated bbt curve while the gray region represents the 95% confidence band. Figure 3 shows a plot for the population and subject-specific curves based on the estimates from the RVM procedure. The thick black curve represents the population average bbt curve while the thin gray curves represent the estimated subject-specific bbt curves. The interval that is characterised with a gentle rise in bbt curve is believed to be the most probable period in the menstrual cycle when the majority of women experience ovulation.

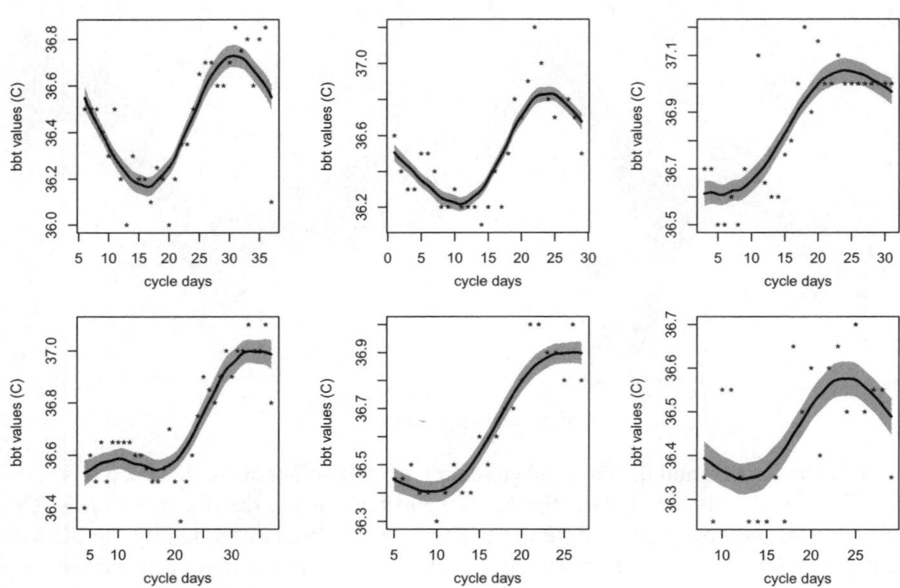

Fig. 2. Estimated bbt curves and the 95% confidence band from the RVM procedure

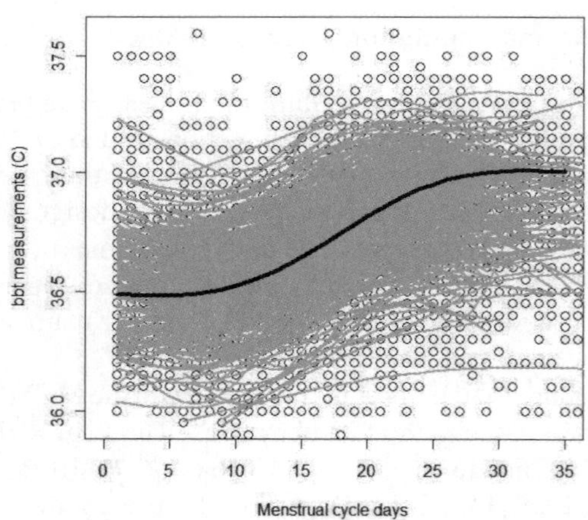

Fig. 3. Estimated population and subject-specific bbt curves from the RVM procedure

3.3 Prediction

To evaluate the predictive ability of the proposed RVM procedure, we conducted an out-of-sample prediction. Typically, we dropped about 20% of the total observations chosen at random from each woman and predicted their values based on a model that contained the remaining bbt measurements. The correlation value between the predicted 20% that were dropped and their corresponding fitted values when all observations are present was 0.80.

Figure 4 shows the predicted bbt values from four randomly selected subjects based on the RVM procedure. The thick line and the gray region in each plot represent the estimated curve and the 95% confidence band based on all observations. The small vertical lines represent the 95% predictive confidence intervals while the star ($*$) at the middle of the vertical lines represents the predicted estimates. This result has substantial clinical implications, as women may be able to collect fewer bbt observations without greatly reducing the accuracy of the estimated bbt curve over the cycle.

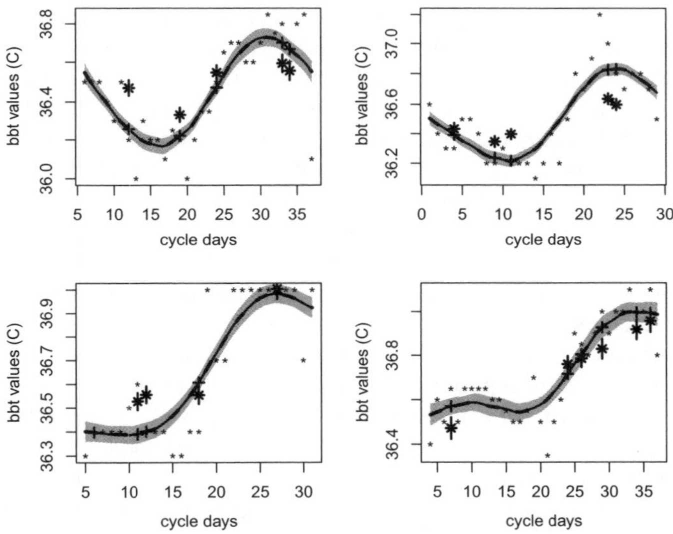

Fig. 4. Estimated bbt curves based on the RVM method and the predicted 20% out of sample bbt values

4 Conclusions

In the literature data smoothing procedures have been used to estimate non-linear curves. For example, [2] used a penalised smoothing spline mixed-model to generate smooth curves for multi-level data. To reduce the dimension of the data [7] used Functional Principal Component Analysis (FPCA) on linear mixed models to generate sparse models that can approximate non-linear curves. However, these MCMC based approaches do not allow fast computation since the computation relies on slow and computational intensive MCMC algorithms. The use of the RVM method can help hasten the computation process leading to a faster model building process that can simultaneously address the two variable selection problems at a less computational cost.

The multi-task Relevant Vector Machine (MT-RVM) approach has been used in machine learning especially in signal reconstruction and compressive sensing applications [13]. However, the application of this approach has not featured in applications that involve smoothing non-linear curves using penalised spline basis functions. In this paper we demonstrate the use of RVM as an alternative approach to computer intensive methods that rely on MCMC. The method is fast and can be used in large dimensional functional models to generate sparse linear and linear mixed models that can be used to rapidly estimate non-linear curves from massive datasets. In this application the RVM procedure was used to smoothen non-linear bbt curves that feature commonly in many reproductive studies.

In linear models there exists numerous variable selection procedures where variable selection procedure can be based on likelihood ratio tests, goodness-of-fit criteria and other methods that are commonly applied in linear regression models. However, in a linear mixed model framework, the variable selection procedures for the random effects normally fail due to complications that arise since the null hypothesis lies on the boundary of the parameter space. Hence, the likelihood ratio (LR) test statistic is no longer valid. This work proposes a fast approximate Bayes approach for simultaneous selection of both fixed and random effects to fit a linear mixed effects model. In adapting the RVM method to perform variable selection, we considered a linear mixed model that assumes independent random effects. This is not always possible in many practical situations; hence [5] have extended the use of RVM methodology into linear mixed models with correlated random effects.

The advantage of our approach is not only computational speed, but it also allows for better generalised performance which can lead to sparse generalised linear mixed models. This aspect can provide inference for a wide variety of models at a moderate computational cost. For example, the RVM methodology can be extended to accommodate probit models as well as handle functional models that have multiple predictors. Moreover, the approach can easily be implemented into more complex and high dimensional functional models with hierarchical data where a subject is allowed to have data from multiple cycles.

References

1. Bigelow, J.L., Dunson, D.B.: Bayesian adaptive regression splines for hierarchical data. Biometrics **63**, 724–732 (2008)
2. Brumback, B.A., Rice, J.A.: Smoothing spline models for the analysis of nested and crossed samples of curves (with discussion). Journal of the American Statistical Association **93**, 961–94 (1998)
3. Burges, C.J.: A Tutorial on Support Vector Machines for Pattern Recognition. Kluwer Academic Publishers, Boston (1998)

4. Ciera, J.M., Scarpa, B., Dunson, D.B.: Fast Bayesian Functional Data Analysis: Application to basal body temperature data. Working Paper Series, Department of Statistical Sciences, University of Padova, Padova (2009)
5. Ciera, J.M., Dunson, D.B.: Fast approximate bayesian functional mixed effects model. Biometrics, to appear (2010)
6. Colombo, B., Masarotto, G.: Daily fecundability: First results from a new data base. Demographic Research **3**(5), (2000)
7. Crainiceanu, C.M.: Bayesian Functional Data Analysis using WinBUGS. Journal of Statistical Software, to appear (2009)
8. Durban, M., Harzelak, J., Wand, M., Carroll, R.: Simple fitting of subject-specifc curves for longitudial data. Statistics in Medicine **24**, 1153–1167 (2005)
9. Hastie, T., Tibshirani, R., Friedman, J.: The Elements of Statistical Learning; Data mining, Inference and Prediction. Springer-Verlag, New York (2001)
10. Green, P.J.: Reversible jump Markov chain Monte Carlo computation and Bayesian model determination. Biometrika **82**, 711–732 (1995)
11. Guo, W.S.: Functional data analysis in longitudinal settings using smoothing splines. Statistical Methods in Medical Research **13**, 49–62 (2004)
12. James, G., Hastie, T., Sugar, C.: Principal component models for sparse functional data. Biometrika **87**, 587–602 (2001)
13. Ji, S., Dunson, D.B., Carin, L.: Multi-task compressive sensing. IEEE Transactions on Signal Processing **57**(1), 92–106 (2009)
14. Jiang, J., Rao, J.S., Gu, Z., Nguyen, T.: Fence methods for mixed model selection. The Annals of Statistics **36**, 1669–1692 (2008)
15. Marshall, J.A.: Field trial of the basal body-temperature method of regulating births. The Lancet **2**, 810 (1968)
16. Pauler, D.K., Wakefield, J.C., Kass, R.E.: Bayes factors and approximations for variance component models. Journal of the American Statistical Association **94**, 1242–1253 (1999)
17. Ramsay, J.O., Silverman, B.W.: Functional data analysis. Springer-Verlag, New York (2005)
18. Rice, J., Wu, C.: Nonparametric mixed effects models for unequally sampled noisy curves. Biometrics **57**, 253–269 (2001)
19. Ruppert, D., Wand, M.P., Carroll, R.J.: Semi-parametric Regression. Cambridge University Press, Cambridge (2003)
20. Scarpa, B., Dunson, D.B.: Bayesian hierarchical functional data analysis via contaminated informative priors. Biometrics **65**(3), 772–780 (2009)
21. Smith, M., Kohn, R.: Nonparametric regression via Bayesian variable selection. Journal of Econometrics **75**, 317–344 (1996)
22. Thompson, W.K., Rosen, O.: A Bayesian model for sparse functional data. Biometrics **64**, 54–63 (2008)
23. Tipping, M.E.: Sparse Bayesian learning and the relevance vector machine. Journal of Machine Learning Research **1**, 211–244 (2001)
24. Wand, M.P., Ormerod, J.T.: On O'Sullivan penalised splines and semiparametric regression. Australian and New Zealand Journal of Statistics **50**, 179–198 (2008)
25. Yao F., Muller H.G., Wang J.L.: Functional data analysis for sparse longitudinal data. Journal of the American Statistical Association **100**(470), 577–591 (2005)
26. Zhang, D., Lin, X., Raz, J., Sowers, M.: Semiparametric stochastic mixed models for longitudinal data. Journal of the American Statistical Association **93**, 710–719 (1998)

A parametric Markov chain to model age- and state-dependent wear processes

Massimiliano Giorgio, Maurizio Guida and Gianpaolo Pulcini

Abstract. Many technological systems are subjected, during their operating life, to a gradual wear process which, in the long run, may cause failure. According to the literature, it results that statisticians and engineers have almost always modeled wear processes by *independent* increments models, which imply that future wear is assumed to depend, at most, on the system's age. In many cases it seems to be more realistic and appropriate to adopt stochastic models which assume that factors other than age affect wear. Indeed, wear models which can (also) account for the dependence on the system's state have been previously proposed in the literature [1, 3, 11, 13]. Many of the abovementioned models present a very complex structure that prevents their application to the kind of data that are usually available. As such, in this paper, a new simple parametric Markov chain wear model is proposed, in which the transition probabilities between process states depend on both the current age and the current wear level of the system. An application based on a real data set referring to the wear process of the cylinder liners of heavy-duty diesel engines for marine propulsion is analysed and discussed.

Key words: wear processes, age and state-dependent wear growth, non-homogeneous Markov chain, negative binomial distribution

1 Introduction

The wear processes of technological units are usually affected by two kind of variability: unit to unit variability [7] and temporal variability [10]. Unit to unit variability determines heterogeneity among the wear paths of different items, which goes beyond what can be accounted for by adopting explanatory variables. This form of variability is usually modeled by introducing unit-specific random effects (see, for example, [7, 9]). Temporal variability, instead, determines the random fluctuation over time of the wear rate of each single unit. This form of variability is usually described via appropriate stochastic processes. This paper is focused on stochastic models which can be used to describe the temporal variability of a (monotonically increasing) wear process.

Since the 1960s, a large body of literature has addressed this problem and many different stochastic models have been proposed (see, for example [1, 2, 4, 11, 12]).

Mantovan, P., Secchi, P. (Eds.): Complex Data Modeling and Computationally Intensive Statistical Methods

Most of the above mentioned literature concentrated its attention on stochastic processes with independent increments, which assume that transition probabilities between process states depend on time (at most), but do not depend on the current process state (*independent increment* assumption). The main advantage in using processes with independent increments is that they often allow for simple solutions to statistical problems such as parameter estimation, hypothesis testing, goodness-of-fit and prediction. Indeed, in the case of many wear processes, the use of models with independent increments does not appear to be fully convincing. In fact, very often it seems more realistic to assume that wear rate can depend on item state. This idea is not new. Wear models which can (also) account for the dependence on the system state were previously proposed in the literature [1, 3, 11, 13]. Nonetheless, many of the abovementioned wear models have never been applied to solve practical problems. Ever better, in almost all the cases in which an application was proposed, only special cases which degenerate in processes with independent increments were effectively applied. As far as we are aware, the only age- and state-dependent wear model proposed up to now in the literature and suitable, at least in principle, for practical applications is the non-homogeneous Markov chain proposed in [1]. A fully operative characterisation of the wear model and estimation procedures were given by the authors only for a simplified version of the proposed model (*i.e.*, a homogeneous Markov chain version), which is able to describe a pure state-dependent wear process only. In addition, the usefulness of the proposed approach appears very limited since the proposed estimation procedure requires, in practice, that the wear level of at least 10 copies of the system under study have to be continuously monitored, until they reach a given wear level of interest. In addition, the model proposed in [1] is non-parametric, thus it cannot be used to describe and/or to predict wear growth beyond the largest observed wear level.

The analysis of the state of the art stimulated a research activity whose aim was to formulate an age- and state-dependent wear model, simple enough to be applied to data actually available in technological applications, which very often arise from uncommonly truncated processes and are obtained via periodic or staggered inspections. As a first partial attempt to fill the abovementioned methodological gap and, in particular, to overcome the limitations of the model proposed in [1], a 3-parameter Markov chain model which is able to describe a pure state-dependent wear process was recently proposed and applied to a set of real wear data [5]. As an additional contribution, in the present paper, a new 4-parameter non-homogeneous Markov chain is proposed, which can be used to model both the dependence on current age and current state. An important feature of the proposed model (hereinafter called the A&SD model) is that a pure state-dependent model and a pure age-dependent model are nested within it. These nested models are themselves valid alternatives to the 3-parameter model presented in [5], and to the widely adopted gamma process [2, 12].

The A&SD model is applied to the case of the degradation process of cylinder liners of identical heavy-duty diesel engines for marine propulsion, which equip three identical ships of the Grimaldi Group. Some preliminary considerations, both of technological and statistical nature, are made to partially motivate the structure of the model. Maximum likelihood estimates of the A&SD model parameters were performed on the basis of real wear measures obtained via staggered inspections. Finally, a formal likelihood ratio test was formulated and applied, which enables one to check whether the wear process under study can be reduced to either the pure age- or pure state-dependent processes nested within the proposed general A&SD model.

2 System description and preliminary technological considerations

Cylinder liner wear is one of the main factors causing failure of heavy-duty marine diesel engines. The maximum wear usually occurs at the Top Dead Centre (TDC) of the liner (see Figure 1), where maximum mechanical and thermal loads are concentrated [8].

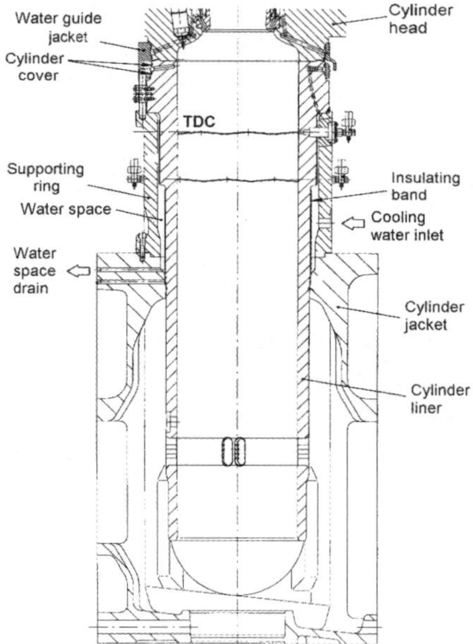

Fig. 1. A cross-section of the 8-cylinder liner of a SULZER RTA 58 engine

Many studies agree that wear at the TDC is mainly due to the presence of a high quantity of abrasive particles on the piston surface, produced by the combustion of heavy fuels and by oil degradation (soot). The wear mechanism may be envisaged as a three-body contact mode. Abrasive contact between the soot and the liner metal surface takes place if the lubricant film results thicker than the soot particle size. Indeed, soot particles are harder than the corresponding engine parts. In addition to abrasive wear, corrosive wear also occurs due to the action of sulfuric acid, nitrous/nitric acids and water. Thus, liner wear can be viewed as a monotonically increasing cumulative damage process. Moreover, since wear increments are (mostly if not entirely) generated by contact between the soot and the liner metal surface, it seems reasonable to assume that both the probability that contact occurs, and the probability that contact produces a given effect, depend on the state of the metal surface. Thus, a model in which the increments depend (also) on state seems to be a good candidate for describing the wear process under study.

3 Data description and preliminary statistical considerations

The data set analysed in this paper consists of 64 measures of the wear level of the TDC observed in $m = 32$ cylinder liners of 8-cylinder SULZER engines. These engines equip a fleet of three identical cargo ships of the Grimaldi Group, which operate on the same routes under homogeneous operating conditions. Data were collected from January 1999 to August 2006 via ad hoc inspections performed on average every 8,000–10,000 hours. Inspection intervals rarely coincide, because the maintenance of propulsion marine engines consists mainly of staggered activities. Moreover, many inspection intervals result larger than expected due to the presence of missing data. Available data are reported in Table 1, where:

- $n_i (i = 1, 2, \ldots, 32)$ denotes the number of inspections performed on liner i;
- $t_{i,j} (i = 1, 2, \ldots, 32;\ j = 1, \ldots, n_i)$ denotes the age (in operating hours) of the liner i at the time of the j-th inspection;
- $w_{i,j} \equiv W\left(t_{i,j}\right)$ denotes the wear (in mm) measured on liner i during the j-th inspection.

Each measure was performed adopting a caliper of sensitivity 0.05 mm. Thus, all the available wear measures are rounded to the nearest multiple of 0.05.

Figure 2 shows the measured wear accumulated by the 32 liners, where the measured data points have been linearly interpolated for graphical representation.

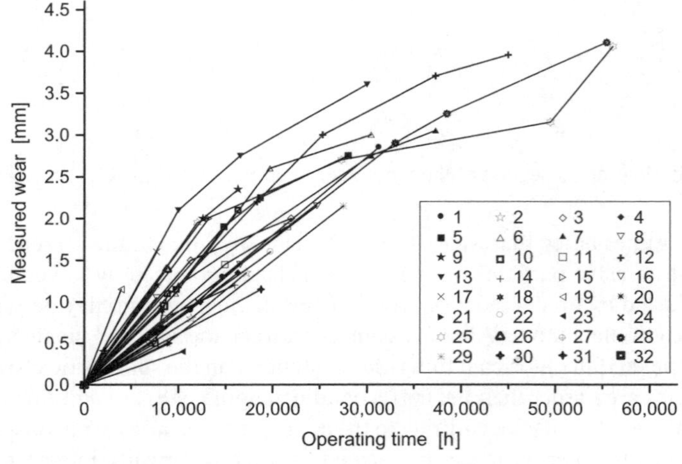

Fig. 2. Observed (interpolated) paths of liner wear

Table 1. Wear data of 32 cylinder liners (t [hours], w [mm])

i	n_i	$t_{i,1}$	$w_{i,1}$	$t_{i,2}$	$w_{i,2}$	$t_{i,3}$	$w_{i,3}$	$t_{i,4}$	$w_{i,4}$
1	3	11,300	0.90	14,680	1.30	31,270	2.85		
2	2	11,360	0.80	17,200	1.35				
3	2	11,300	1.50	21,970	2.00				
4	2	12,300	1.00	16,300	1.35				
5	3	14,810	1.90	18,700	2.25	28,000	2.75		
6	3	9,700	1.10	19,710	2.60	30,450	3.00		
7	3	10,000	1.20	30,450	2.75	37,310	3.05		
8	3	6,860	0.50	17,200	1.45	24,710	2.15		
9	3	2,040	0.40	12,580	2.00	16,620	2.35		
10	4	7,540	0.50	8,840	1.10	9,770	1.15	16,300	2.10
11	3	8,510	0.80	14,930	1.45	21,560	1.90		
12	4	18,320	2.20	25,310	3.00	37,310	3.70	45,000	3.95
13	3	10,000	2.10	16,620	2.75	30,000	3.60		
14	2	9,350	0.85	15,970	1.20				
15	1	13,200	2.00						
16	1	7,700	1.05						
17	1	7,700	1.60						
18	1	8,250	0.90						
19	1	3,900	1.15						
20	1	7,700	1.20						
21	1	9,096	0.50						
22	1	19,800	1.60						
23	1	10,450	0.40						
24	1	12,100	1.00						
25	4	12,000	1.95	27,300	2.70	49,500	3.15	56,120	4.05
26	1	8,800	1.40						
27	1	2,200	0.40						
28	3	33,000	2.90	38,500	3.25	55,460	4.10		
29	2	8,800	0.50	27,500	2.15				
30	1	8,250	0.70						
31	1	18,755	1.15						
32	1	8,490	0.95						

Observing the wear paths, it is possible to recognise an accommodation period, with an overall decreasing wear rate, which occupies almost the entire observation period, a behaviour that is in agreement with the literature (see, for example, [4]). Moreover, the observed (interpolated) paths are closely interwoven, the rate of wear of an individual liner changes with operating time, and these changes are of a random nature. Thus, this application shows an evident presence of temporal variability. On the contrary, this preliminary study gives no clear evidence in favour of the presence of

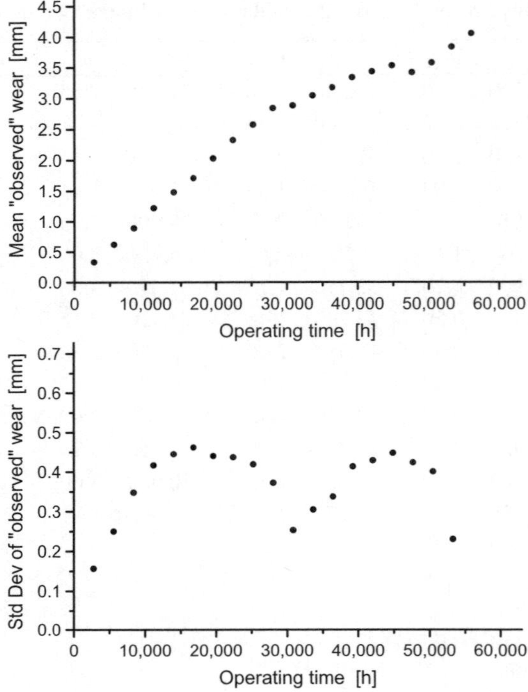

Fig. 3. Empirical estimates of the mean and standard deviation of liner wear

unit to unit variation. Although these visual analyses cannot be considered conclusive, only the temporal variability will be taken into account in the following discussion. A more comprehensive study concerning the possible presence of random effects is presented in [5].

Figure 3 shows the empirical estimates of the mean and standard deviation of the wear process as a function of time. Since the inspection times generally differ, and thus wear measures generally refer to different ages of the liners, these empirical estimates were obtained by using an interpolation procedure at selected equispaced times $\tau_k = k \cdot \max_i(t_{i,n_i})/20$ ($k = 1, 2, \ldots, 20$). Namely, for each liner i such that $t_{i,n_i} \geq \tau_k$, the linearly interpolated wear value, say $W_i(\tau_k)$, is obtained by

$$W_i(\tau_k) = \frac{W(t_{i,j}) - W(t_{i,j-1})}{t_{i,j} - t_{i,j-1}} \cdot (\tau_k - t_{i,j-1}) + W(t_{i,j-1}); \qquad i : t_{i,n_i} \geq \tau_k,$$

where $t_{i,j}$ is the smallest inspection time of liner i larger than or equal to τ_k, and m_k is the number of liners whose last inspection time is larger than or equal to τ_k. Then, the empirical estimates of the mean and standard deviation are obtained as:

$$\tilde{E}\left(W(\tau_k)\right) = \frac{\sum\limits_{i:t_{i,n_i} \geq \tau_k} W_i(\tau_k)}{m_k}$$

$$\text{St}\tilde{\text{d}}\text{Dev}\left(W(\tau_k)\right) = \left(\frac{\sum_{i:t_{i,n_i} \geq \tau_k} \left[W_i(\tau_k) - \tilde{E}\left(W(\tau_k)\right) \right]^2}{m_k - 1} \right)^{1/2}.$$

Figure 3 clearly shows that the mean wear is a downward (monotone increasing) function of time and, even more important, the estimated standard deviation presents a non-monotone trend. This behaviour cannot be reproduced by a model with independent increments, whose standard deviation increases monotonically with system age. Obviously, this explorative analysis is not considered to give strong evidence against the hypothesis that the wear process under study has independent increments. Nevertheless, it constitutes a considerable clue in favour of the idea that, in this application, an adequate wear model is one which assumes that factors other than system age are affecting wear.

4 Model description

Let $\{W(t); t \geq 0\}$ denote the (non-decreasing, continuous time, continuous state) process that describes the evolution of wear accumulated by an item over the operating time t, when starting from a known initial state $W(0) = W_0$. By dividing the time and wear axes into contiguous and equispaced intervals of length h and Δ respectively, a discrete time, discrete state stochastic model, say $\{W(\upsilon); \upsilon = 0, 1, 2 \ldots\}$, which approximates the actual wear process is derived, where $W(\nu)$ represents the system state at time $t_\nu = \nu h$. This model is assumed to be a non-homogeneous Markov chain with countable state space $\{W(\nu) = k \cdot \Delta; k = 0, 1, 2, \ldots\}$.

The elements $[\boldsymbol{P}(\nu)]_{k,k+s}$ $(k, s = 0, 1, 2, \ldots)$ of the one–step transition matrix $\boldsymbol{P}(\nu)$ are defined by

$$\begin{aligned}
[\boldsymbol{P}(\nu)]_{k,k+s} &\equiv \Pr\{W(\nu+1) = (k+s)\Delta | W(\nu) = k\Delta\} = \\
&= \binom{s + \psi(\nu; a, b) - 1}{\psi(\nu; a, b) - 1} \left[\frac{\Delta}{\varphi(k; c, d) + \Delta} \right]^{\psi(\nu;a,b)} \left[\frac{\varphi(k; c, d)}{\varphi(k; c, d) + \Delta} \right]^s, \\
&\quad s = 0, 1, 2, \ldots
\end{aligned} \tag{1}$$

where $\psi(\nu; a, b)$ and $\phi(k; c, d)$ are two non-negative functions that account for the dependence on age and state respectively. These functions have to be properly defined in order to well describe the wear process under study. In this paper, the commonly used power law function

$$\psi(\nu; a, b) = [(\nu + 1)h/a]^b - (\nu h/a)^b, \quad a, b > 0$$

has been used to account for the dependence on age, and the classical exponential smoothing function

$$\phi(k; c, d) = d \exp(-c \cdot k \cdot \Delta), \quad d > 0, -\infty < c < +\infty$$

has been adopted to account for the dependence on state, due to its flexibility. In fact, the exponential function allows a description of constant, monotone increasing and monotone decreasing behaviours, depending on the value of the shape parameter c.

Thus, according to model (1), the transition probability $[\boldsymbol{P}(\nu)]_{k,k+s}$ depends both on the current age $t = \nu \cdot h$, and on the current state $W(\nu) = k \cdot \Delta$ of the system. Moreover, model (1) assumes that, given the system wear level $k \cdot \Delta$ at the age $t_\nu = \nu h$, the (discrete) wear increment $W(\nu + 1) - W(\nu)$ over the elementary time interval $[\nu h, (\nu + 1)h]$ has a negative binomial distribution with mean

$$\mathrm{E}(W(\nu + 1) - W(\nu)|W(\nu) = k \cdot \Delta) = \psi(\nu; a, b)\phi(k; c, d)$$

and variance

$$\mathrm{Var}(W(\nu + 1) - W(\nu)|W(\nu) = k \cdot \Delta) = \psi(\nu; a, b)\phi(k; c, d)[\phi(k; c, d) + \Delta].$$

From (1), the δ–step ($\delta > 1$) transition matrix, say $\boldsymbol{P}^{(\delta)}(\nu)$, whose element $[\boldsymbol{P}^{(\delta)}(\nu)]_{k,k+s}$ gives the transition probability from state k to state $k + s$ during the time interval $(kh, (k + s)h)$, is given by

$$\boldsymbol{P}^{(\delta)}(\nu) = \prod_{r=\nu}^{\nu+\delta-1} \boldsymbol{P}(r). \tag{2}$$

From (2), the mean and variance of the wear level at time $t_\delta = \delta h$, given the initial state $W(0) = k_0 \Delta$, are, respectively:

$$E(W(\delta)|k_0) = \sum_{s=0}^{\infty} (k_0 + s)\Delta \cdot [\boldsymbol{P}^{(\delta)}(0)]_{k_0,k_0+s}, \quad \delta = 1, 2, \ldots \tag{3}$$

$$\mathrm{Var}(W(\delta)|k_0) = \sum_{s=0}^{\infty} [(k_0 + s)\Delta - E(W(\delta)|k_0)]^2 \cdot [\boldsymbol{P}^{(\delta)}(0)]_{k_0,k_0+s}; \quad \delta = 1, 2, \ldots . \tag{4}$$

Model (1) includes, as a special case: a) a pure age-dependent model, b) a pure state-dependent model, and c) a homogeneous process with independent increments. Indeed, when $c = 0$, the function $\phi(k; c, d)$ becomes a constant independent of the current state and the A&SD model (1) is reduced to a pure age-dependent (AD) Markov chain with negative binomial independent increments. Due to the relationship between negative binomial and gamma random variables [6], this AD model can be regarded as the discrete time-discrete state version of the widely applied gamma process [2,12]. On the other hand, when $b = 1$, the function $\psi(\nu; a, b)$ becomes a constant independent of the current age, and the A&SD model (1) is reduced to a pure state-dependent (SD) homogeneous Markov chain, which is practically equivalent to the SD model presented in [5]. Finally, when $c = 0$ and $b = 1$, the A&SD model (1) is reduced to a homogeneous, independent increment Markov process which can be considered as the discrete state version of gamma process with independent and stationary increments.

Note that, when $c = 0$, the δ–step transition matrix no longer requires the product of matrices, is independent of the state k, and can also be used for non-integer values of δ. In particular, for the AD model

$$[\boldsymbol{P}^{(\delta)}(\nu)]_{k,k+s} = \binom{s + \psi^{(\delta)}(\nu; a, b) - 1}{\psi^{(\delta)}(\nu; a, b) - 1} \left[\frac{\Delta}{d+\Delta}\right]^{\psi^{(\delta)}(\nu;a,b)} \left[\frac{d}{d+\Delta}\right]^{s}, \quad s = 0, 1, 2, \ldots,$$

where $\psi^{(\delta)}(\nu; a, b) = [(\nu + \delta)h/a]^b - (\nu h/a)^b$, whereas for the homogeneous, independent increments Markov process:

$$[\boldsymbol{P}^{(\delta)}(\nu)]_{k,k+s} = \binom{s + (\delta h/a) - 1}{(\delta h/a) - 1} \left[\frac{\Delta}{d+\Delta}\right]^{(\delta h/a)} \left[\frac{d}{d+\Delta}\right]^{s}, s = 0, 1, 2, \ldots .$$

5 Parameter estimation

Let $\upsilon_{i,j} = int(t_{i,j}/h + 0.5)$ denote the age of the liner i ($i = 1, \ldots, m$) at the jth inspection epoch ($j = 1, \ldots, n_i$) in the discretised time axis, and let $k_{i,j} = w_{i,j}/\Delta$ denote the state of the liner i at the jth inspection epoch within the discrete–state approximation. Note that, in the present application wear measures are rounded to the nearest multiple of 0.05 mm and Δ is chosen exactly equal to 0.05, then each ratio $w_{i,j}/\Delta$ is an integer.

From (2), the probability of the wear $W(\upsilon_{i,j})$ accumulated by the liner i up until inspection time $\upsilon_{i,j} \cdot h \cong t_{i,j}$, given the state $W(\upsilon_{i,j-1}) = k_{i,j-1}\Delta$ at the previous inspection time $\upsilon_{i,j-1} \cdot h \cong t_{i,j-1}$, results in

$$\Pr\{W(\nu_{i,j}) = k_{i,j}\,\Delta | W(\nu_{i,j-1}) = k_{i,j-1}\,\Delta\} = \left[\prod_{r=\nu_{i,j-1}}^{\upsilon_{i,j}-1} \boldsymbol{P}(r)\right]_{k_{i,j-1},k_{i,j}},$$

where $t_{i,0} \equiv 0$, $\nu_{i,0} \equiv 0$, and $k_{i,0} \equiv 0$, and the approximation arises because $\nu_{i,j} \cong t_{i,j}/h$ and $\nu_{i,j-1} \cong t_{i,j-1}/h$. Then, since the probability distribution of $W(\nu_{i,j})$ depends on the history only through age and state of the liner at the previous inspection, the joint probability of the observed wear levels of each liner i is given by

$$\prod_{j=1}^{n_i} \Pr\{W(\nu_{i,j}) = k_{i,j} | W(\nu_{i,j-1}) = k_{i,j-1}\}.$$

Thus, the likelihood function is

$$L(a, b, c, d|\text{data}) = \prod_{i=1}^{m} \prod_{j=1}^{n_i} \left[\prod_{r=\nu_{i,j-1}}^{\nu_{i,j}-1} \boldsymbol{P}(r)\right]_{k_{i,j-1},k_{i,j}}, \tag{5}$$

where the elements of $\boldsymbol{P}(r)$ are implicitly functions of the model parameters a, b, c and d through (1). These parameters are estimated by maximising the logarithm of (5) through a numerical optimisation procedure. Since $\boldsymbol{P}(r)$ is, in principle, an infinite matrix, we treated the state $k_{\max}+1$, where $k_{\max}$ is the largest observed wear state ($k_{\max} = 82$ in the present application), as an absorbing state, and we replaced the matrix $\boldsymbol{P}(r)$ by the finite $(k_{\max}+2 \times k_{\max}+2)$ matrix $\boldsymbol{Q}(r)$. By setting $h = 500$ hours and $\Delta = 0.05$mm, the ML estimates of a, b, c and d are

$$\hat{a} = 1379.7\text{h}, \quad \hat{b} = 1.021, \quad \hat{c} = -0.3162\text{mm}^{-1}, \quad \hat{d} = 0.1823\text{mm},$$

and the ML estimate, say $\hat{\boldsymbol{P}}(r)$, of the one-step transition matrix $\boldsymbol{P}(r)$ is easily obtained from (1). ML estimates of mean and variance of the wear level at the time $t_\delta = \delta h$, given the initial state $W(0) = k_0\Delta$, are obtained by substituting $\hat{\boldsymbol{P}}(r)$ in place of $\boldsymbol{P}(r)$ in expressions (3) and (4) respectively. The ML estimates of mean and standard deviation are depicted in Figure 4, together with the empirical estimates of Figure 3. The plots show that the ML estimate of the mean wear (3) follows the observed data very well, and that the ML estimate of the standard deviation (4) reproduces the non-monotone behaviour of the empirical estimate.

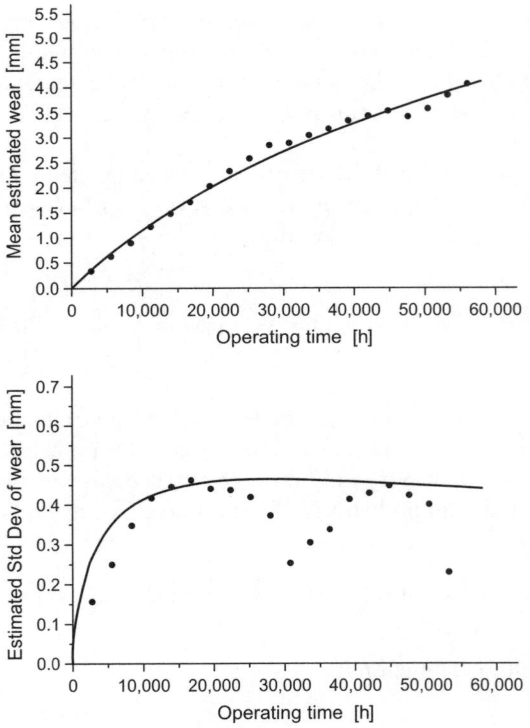

Fig. 4. Empirical estimate (•) and ML estimate under the A&SD model (continuous lines) of mean and standard deviation of liner wear

6 Testing dependence on time and/or state

Let: θ be the $p \times 1$ vector of model parameters, and Ω the p-dimensional parameter space; $H_i = \{\theta : \theta \in \Omega_i\}$ $(i = 0, 1)$ the hypothesis set on θ, where Ω_0 is the parameter space when r $(r < p)$ components of θ are perfectly known, and $\Omega_1 = \Omega - \Omega_0$. It is well known that, in repeated sampling, the statistic

$$\Lambda_r = -2 \ln \left[\sup_{\theta:\theta\in\Omega} L(\theta|\text{data}) - \sup_{\theta:\theta\in\Omega_0} L(\theta|\text{data}) \right] \tag{6}$$

is approximately distributed as a chi-square random variable with r degrees of freedom.

Based on the statistic (6), three different null hypotheses were tested against the alternative hypothesis H_1 of a non-homogeneous Markov chain (1) with state dependent increments, namely:

- a homogeneous Markov chain with independent increments, *i.e.*, $b = 1$ and $c = 0$;
- a non-homogeneous Markov chain with independent increments (AD model), *i.e.*, $c = 0$;
- a homogeneous Markov chain with state-dependent increments (SD model), *i.e.*, $b = 1$.

From the estimation results given in Table 2 and the estimated log-likelihood under the alternative hypothesis H_1, say $\ln L(\hat{a}, \hat{b}, \hat{c}, \hat{d}|\text{data}) = -213.158$, we have that:

- the null hypothesis of a homogeneous Markov chain with independent increments is rejected ($\Lambda_2 = 27.31$, p–value = 1.2×10^{-6});
- the null hypothesis of an AD model is rejected ($\Lambda_1 = 7.46$, p–value = 0.0063);
- the null hypothesis of an SD model is not rejected ($\Lambda_1 = 0.021$, p–value = 0.886).

Thus, it is reasonable to assume that the factor which mainly affects the wear process of cylinder liners is the current liner state, not the liner age.

Table 2. Estimation results under the null hypotheses

Null hypothesis	$\hat{a}$[h]	$\hat{b}$	$\hat{c}$[mm^{-1}]	$\hat{d}$[mm]	*Estimated log-likelihood*
Markov chain indep. increments	1646.0	1	0	0.1544	−226.814
AD model	559.4	0.8138	0	0.1056	−216.887
SD model	1283.2	1	0.2977	0.1756	−213.168

Figure 5 compares the estimated mean and standard deviation curves within the A&SD model to the corresponding curves obtained within the SD model, the AD model, and the homogeneous chain with independent increments. The estimated curves show that the SD model fits the observed wear process as well as the A&SD

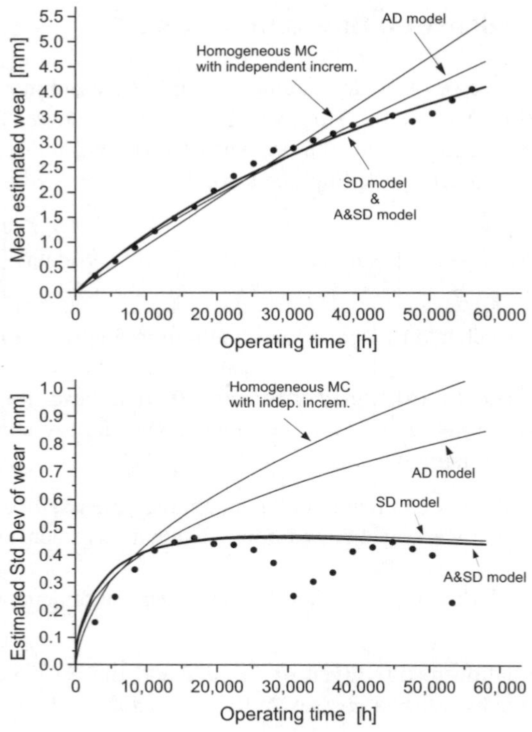

Fig. 5. Empirical (•) and ML (continuous lines) estimates of mean and standard deviation of liner wear

model (in particular, the mean curves within these two models are indistinguishable and the standard deviation curves are very close to each other), whereas both the AD model and, moreover, the homogeneous Markov chain with independent increments are inadequate to describe the wear process.

7 Conclusions

In this paper an innovative degradation model is proposed, in which the transition probabilities between states depend both on the current age and current state of the system. The proposed model is expected to be suitable for describing a wide class of degradation phenomena since it can be easily applied also to pure age-dependent and pure state-dependent processes. In particular, the proposed model appears to fit well the wear data obtained from cylinder liners of diesel engines which equip three identical ships of the Grimaldi Group. Performing testing procedures, based on the likelihood ratio statistic, allows one to assess whether the degradation process under study really depends on the age and/or state of the system. In particular, the wear process of cylinder liners is shown to be strongly affected by the system state, not by its age.

Acknowledgement. The present work was developed with the contribution of the Italian Ministry of Education, University and Research (MIUR) within the framework of the PRIN-2005 project: "Statistical design of 'continuous' product innovation".

References

1. Bogdanoff, J.L., Kozin, F.: Probabilistic Models of Cumulative Damage. John Wiley & Sons, New York (1985)
2. Çinlar, E.: On a generalization of Gamma process. Journal of Applied Probability **17**, 467–480 (1980)
3. Durham, S.D., Padgett, W.J.: Cumulative damage models for system failure with application to carbon fibers and composites. Technometrics **39**, 34–44 (1997)
4. Gertsbakh, I.B., Kordonskiy, K.B.: Models of Failure. Springer-Verlag, Berlin (1969)
5. Giorgio, M., Guida, M., Pulcini, G.: A state-dependent wear model with an application to marine engine cylinder liners. Technometrics, in press (DOI: 10.1198/TECH.2009.08092) (2010)
6. Kozubowski, T.J., Podgorski, K.: Invariance properties of the negative binomial Lèvy process and stochastic self-similarity. International Mathematical Forum **2**, 1457–1468 (2007)
7. Lawless, J., Crowder, M.: Covariates and random effects in a gamma process model with application to degradation and failure. Lifetime Data Analysis **10**, 213–227 (2004)
8. Li, S., Csontos, A.A., Gable, B.M., Passut, C.A., Jao, T.C.: Wear in Cummins M-11/EGR test engine. In: *Proc. SAE International Fuels & Lubricants Meeting & Exhibition*. Reno, NV, Society of Automotive Engineers, SAE Paper 2002-01-1672 (2002)
9. Meeker, W.Q., Escobar, L.A.: Statistical Methods for Reliability Data. John Wiley & Sons, New York (1998)
10. Pandey, M.D., van Noortwijk, J.M.: Gamma process model for time-dependent structural reliability analysis. In: Watanabe, E.. Frangopol, D.M., Utsonomiya, T. (eds.) *Bridge Maintenance, Safety, Management and Cost. Proceedings of the Second International Conference on Bridge Maintenance, Safety and Management (IABMAS), Kyoto, Japan, 18–22 October 2004*. Taylor & Francis Group, London (2004)
11. Singpurwalla, N.D.: Survival in dynamic environments. Statistical Science **10**, 86–103 (1995)
12. van Noortwijk, J.M., Kallen, M.J., Pandey, M.D.: Gamma processes for time-dependent reliability of structures. In: Kolowrocki, K. (ed.) *Advances in Safety and Reliability. Proceedings of ESREL 2005 – European Safety and Reliability Conference 2005, Tri City (Gdynia-Sopot-Gdansk), Poland, 27–30 June 2005*. Taylor & Francis Group, London, 1457–1464 (2005)
13. Wenocur, M.L.: A reliability model based on the gamma process and its analytic theory. Advances in Applied Probability **21**, 899–918 (1989)

Case studies in Bayesian computation using INLA

Sara Martino and Håvard Rue

Abstract. Latent Gaussian models are a common construct in statistical applications where a latent Gaussian field, indirectly observed through data, is used to model, for instance, time and space dependence or the smooth effect of covariates. Many well-known statistical models, such as smoothing-spline models, space time models, semiparametric regression, spatial and spatio-temporal models, log-Gaussian Cox models, and geostatistical models are latent Gaussian models. Integrated Nested Laplace approximation (INLA) is a new approach to implement Bayesian inference for such models. It provides approximations of the posterior marginals of the latent variables which are both very accurate and extremely fast to compute. Moreover, INLA treats latent Gaussian models in a general way, thus allowing for a great deal of automation in the inferential procedure. The `inla` programme, bundled in the `R` library `INLA`, is a prototype of such black-box for inference on latent Gaussian models which is both flexible and user-friendly. It is meant to, hopefully, make latent Gaussian models applicable, useful and appealing for a larger class of users.

Key words: approximate Bayesian inference, latent Gaussian model, Laplace approximations, structured additive regression models

1 Introduction

Latent Gaussian models are an apparently simple but very flexible construct in statistical applications which covers a wide range of common statistical models spanning from (generalised) linear models, (generalised) additive models, smoothing spline models, state space models, semiparametric regression, spatial and spatiotemporal models, log-Gaussian Cox processes and geostatistical and geoadditive models. In these models, the latent Gaussian field serves as a flexible and powerful tool to model non-linear effects of covariates, group specific heterogeneity, as well as space and time dependencies among data.

Bayesian inference on latent Gaussian models is not straightforward since, in general, the posterior distribution is not analytically available. Markov Chain Monte Carlo (MCMC) techniques are, today, the standard solution to this problem and several ad hoc algorithms have been developed in recent years. Although in theory always

Mantovan, P., Secchi, P. (Eds.): Complex Data Modeling and Computationally Intensive Statistical Methods

possible to implement, MCMC algorithms applied to latent Gaussian models come with a wide range of problems in terms of convergence and computational time. Moreover, the implementation itself might often be problematic, especially for end users who might not be experts in programming.

Integrated Nested Laplace approximation (INLA) is a new tool for Bayesian inference on latent Gaussian models when the focus is on posterior marginal distributions [20]. INLA substitutes MCMC simulations with accurate, deterministic approximations to posterior marginal distributions. The quality of such approximations is extremely high, such that even very long MCMC runs could not detect any error in them. A detailed description of the INLA method and a thorough comparison with MCMC results can be found in [20].

INLA presents two main advantages over MCMC techniques. The first and most outstanding is computational. Using INLA results are obtained in seconds and minutes even for models with a huge dimensional latent field, while a well build MCMC algorithm would take hours or even days. This is also due to the fact that INLA is naturally parallelised, thus making it possible to exploit the new trend of having multi-core processors. The second, and not less important advantage, is that INLA treats latent Gaussian models in a unified way, thus allowing greater automation of the inference process. The core of the computational machinery, automatically adapts to any kind of latent field so that, from the computational point of view, it does not matter if we deal with, for example, spatial or temporal models. In practice INLA can be used almost as a black box to analyse latent Gaussian models.

A prototype of such programme, `INLA`, together with a user-friendly `R` interface (`INLA` library) is already available from the web-site `www.r-inla.org`. Its goal is to make the INLA approach available for a larger class of users. The hope is that near instant inference and simplicity of use will make latent Gaussian models more applicable, useful and appealing for the end user.

The purpose of this paper is to give an overview of models to which INLA is applicable. We will present a series of case studies ranging from generalised linear models to spatially varying regression models to survival models, solved using the INLA methodology through the `INLA` library. The structure of this article is as follows. Section 2 describes latent Gaussian models and their main features. Section 3 and Section 4 briefly introduce the INLA approach and the `INLA` library. In Section 5 three case studies are analysed. They include a GLMM model with over-dispersion, different models for spatial analysis and a model for survival data. We end with a brief discussion in Section 6.

2 Latent Gaussian models

Latent Gaussian models are hierarchical models which assume an n-dimensional Gaussian field $\mathbf{x} = \{x_i : i \in \mathcal{V}\}$ to be point-wise observed through n_d conditional independent data $\mathbf{y}$. Both the covariance matrix of the Gaussian field $\mathbf{x}$ and the likelihood model for $y_i|\mathbf{x}$ can be controlled by some unknown hyperparameters θ. In addition, some linear constraints of the form $\mathbf{Ax} = \mathbf{e}$, where the matrix $\mathbf{A}$ has rank k, may be imposted. The posterior then reads:

$$\pi(\mathbf{x}, \theta \mid \mathbf{y}) \;\propto\; \pi(\theta)\; \pi(\mathbf{x} \mid \theta) \prod_{i \in \mathcal{I}} \pi(y_i \mid x_i, \theta). \tag{1}$$

As the likelihood is not often Gaussian, this posterior density is not analytically tractable.

A slightly different point of view to look at latent Gaussian models is to consider structured additive regression models; these are a flexible and extensively used class of models, see for example [8] for a detailed account. Here, the observation (or response) variable y_i is assumed to belong to an exponential family where the mean μ_i is linked to a structured additive predictor η_i through a link-function $g(\cdot)$, so that $g(\mu_i) = \eta_i$. The likelihood model can be controlled by some extra hyperparameters θ_1. The structured additive predictor η_i accounts for effects of various covariates in an additive way:

$$\eta_i = \beta_0 + \sum_{j=1}^{n_f} w_{ji} f^{(j)}(u_{ji}) + \sum_{k=1}^{n_\beta} \beta_k z_{ki} + \epsilon_i. \tag{2}$$

Here, the $\{\beta_k\}$'s represent the linear effect of covariates $\mathbf{z}$. The $\{f^{(j)}(\cdot)\}$'s are unknown functions of the covariates $\mathbf{u}$. These can take very many different forms: non-linear effects of continuous covariates, time trends, seasonal effects, i.i.d. random intercepts and slopes, group specific random effects and spatial random effects can all be represented through the $\{f^{(j)}\}$'s functions. The w_{ij} are known weights defined for each observed data point. Finally, ϵ_i's are unstructured random effects.

A latent Gaussian model is obtained by assigning $\mathbf{x} = \{\{f^{(j)}(\cdot)\}, \{\beta_k\}, \{\eta_i\}\}$, a Gaussian prior with precision matrix $\mathbf{Q}(\theta_2)$, with hyperparameters θ_2. Note that we have parametrised the latent Gaussian field so that it includes the η_i's instead of the ϵ_i's, in this way some of the elements of $\mathbf{x}$, namely the η_i's, are observed through the data $\mathbf{y}$. This is mainly due to the fact that the `INLA` library requires each data point y_i to be dependent on the latent Gaussian field only through one single element of $\mathbf{x}$, namely η_i. For this reason, a small random noise, ϵ_i, with high precision is always automatically added to the model. The definition of the latent model is completed by assigning the hyperparameters $\theta = (\theta_1, \theta_2)$ a prior distribution.

In this paper the latent Gaussian models are assumed to satisfy two basic properties: First, the latent Gaussian model $\mathbf{x}$, often of large dimension, admits conditional independence properties. In other words it is a latent Gaussian Markov random field (GMRF) with a sparse precision matrix $\mathbf{Q}$ [18]. The second property is that the dimension m of the hyperparameter vector θ is small, say ≤ 6. These properties are satisfied by many latent Gaussian models in the literature. Exceptions exist, geostatistical models being the main one. INLA can still be applied to geostatistical models using different computational machinery or using a Markov representation of the Gaussian field (see [7] and the discussion contribution from Finn Lindgren in [20]).

3 Integrated Nested Laplace Approximation

Integrated Nested Laplace Approximation (INLA) is a new approach to statistical inference for latent Gaussian models introduced by [19] and [20]. In short, the INLA approach provides a recipe for fast Bayesian inference using accurate approximations of the marginal posterior density for the hyperparameters $\tilde{\pi}(\theta|\mathbf{y})$ and for the full conditional posterior marginal densities for the latent variables $\tilde{\pi}(x_i|\theta, \mathbf{y})$, $i = 1, \ldots, n$. The approximation for $\pi(\theta|\mathbf{y})$ is based on the Laplace approximation [22], while for $\pi(x_i|\theta, \mathbf{y})$ three different approaches are possible: a Gaussian, a full Laplace and a simplified Laplace approximation. Each of these has different features, computing times and accuracy. The Gaussian approximation is the fastest to compute but there can be errors in the location of the posterior mean and/or errors due to the lack of skewness [19]. The Laplace approximation is the most accurate but its computation can be time consuming. Hence, in [20], the simplified Laplace approximation is introduced. This is fast to compute and usually accurate enough.

Posterior marginals for the latent variables $\tilde{\pi}(x_i|\mathbf{y})$ are then computed via numerical integration such as:

$$\begin{aligned}\tilde{\pi}(x_i|\mathbf{y}) &= \int \tilde{\pi}(x_i|\theta, \mathbf{y})\tilde{\pi}(\theta|\mathbf{y})\, d\theta \\ &\approx \sum_{k=1}^{K} \tilde{\pi}(x_i|\theta_k, \mathbf{y})\tilde{\pi}(\theta_k|\mathbf{y})\ \Delta_k\,. \end{aligned} \tag{3}$$

Posterior marginals for the hyperparameters $\tilde{\pi}(\theta_j|\mathbf{y})$, $J = 1, \ldots, m$ are computed in a similar way. The choice of the integration points θ_k can be done using two strategies: the first strategy, more accurate but also time consuming, is a to define a grid of points covering the area where most of the mass of $\tilde{\pi}(\theta|\mathbf{y})$ is located (GRID strategy); the second strategy, named central composit design (CCD strategy), comes from the design problem literature and consists of laying out a small amount of 'points' in an m-dimensional space in order to estimate the curvature of $\tilde{\pi}(\theta|\mathbf{y})$, see [20] for more details on both strategies. In [20] it is suggested to use the CCD strategy as a default choice. Such strategy is usually accurate enough for the computation of $\tilde{\pi}(x_i|\mathbf{y})$, while a GRID strategy might be necessary if one is interested in an accurate estimate of $\tilde{\pi}(\theta_j|\mathbf{y})$. The approximate posterior marginals obtained from such procedure can then be used to compute summary statistics of interest, such as posterior means, variances or quantiles.

INLA can also compute, as a by-product of the main computations, other quantities of interest like Deviance Information Criteria (DIC) [21], marginal likelihoods and predictive measures as logarithmic scores [11] and the PIT histogram [6], useful to detect outliers and to compare and validate models.

Different strategies to assess the accuracy of the various approximations for the densities $x_i|\theta, \mathbf{y}$ are described in [20]. The INLA approximations assume the posterior distribution $\pi(\mathbf{x}|\theta, \mathbf{y})$ to be unimodal and fairly regular. This is usually the case for most real problem and data sets. INLA can, however, deal to some extent with the multimodality of $\pi(\theta|\mathbf{y})$, provided that the modes are sufficiently closed. See the discussion contributions of Ferreira and Hodges and the author's reply in [20].

Theory and practicalities surrounding INLA are extensively analysed in [20] and will not be repeated here. Loosely speaking we can say that INLA fully exploits all the main features of the latent Gaussian models described in Section 2. Firstly, all computations are based on sparse matrix algorithms which are much faster than the corresponding algorithms for dense matrix. Secondly, the presence of the latent Gaussian field and the usual "good behaviour" of the likelihood function justify the accuracy of the Laplace approximation. Finally, the small number of hyperparameters θ makes the numerical integration in equation (3) computationally feasible.

4 The `INLA` package for `R`

Computational speed is one of the most important components of the INLA approach, therefore special care has to be put into the implementation of the required algorithms. All computations required by the INLA methodology are efficiently performed by the `inla` programme, written in `C` based on the `GMRFLib`-library which includes efficient algorithms for sparse matrices [18]. Both the `inla` programme and the `GMRFLib`-library, in addition, use the OpenMP (see `http://www.openmp.org`) to speed up the computations for shared memory machines, i.e. multicore processors, which are today standard for new computers.

Moreover, the `inla` programme is bundled within an `R` library called `INLA` in order to aid its usage. The software is open-source and can be downloaded from the web site `www.r-inla.org`. It is run by Linux, MAC and Windows. On the same website documentation and a large sample of applications are also provided.

5 Case studies

The following examples are meant to give an overview of the range of application of the INLA methodology. All examples are implemented using the `INLA` library on a dual-core 2.5GHz laptop. The `R` code is reported where it was considered helpful. The rest of the `R` code can be downloaded from the `www.r-inla.org` website in the Download section.

5.1 A GLMM with over-dispersion

The first example is a generalised linear mixed model with binomial likelihood, where random effects are used to model within group extra variation. The data concern the proportion of seeds that germinated on each of $m = 21$ plates arranged in a 2×2 factorial design with respect to seed variety and type of root extract. The data set was presented by [5] and analysed among others by [3]. This example is also included in the WinBUGS/OpenBUGS manual [15]. In [9] the authors perform a comparison between the INLA and the maximum likelihood approach for this particular data set.

The sampling model is $y_i|\eta_i \sim$ Binomial (n_i, p_i) where, for plate $i = 1, \ldots, m$, y_i is the number of germinating seeds (variable name (vn): `r`) and n_i the total number of seeds ranging from 4 to 81 (vn: `n`), and $p_i = \text{logit}^{-1}(\eta_i)$ is the unknown probability of germinating. To account for between plate variability, [3] introduce plate-specific random effects, and then fit a model with main and interaction effects:

$$\eta_i = \beta_0 + \beta_1 z_{1i} + \beta_2 z_{2i} + \beta_3 z_{1i} z_{2i} + f(u_i), \tag{4}$$

with z_{1i} and z_{2i} representing the seed variety (vn: `x1`) and type of root extract (vn: `x2`) of plate i. We assign $\beta_k, k = 0, \ldots, 3$ vague Gaussian priors with known precision. Moreover, we assume $f(u_i)|\tau_u \sim \mathcal{N}(0, \tau_u^{-1}), i = 1, \ldots, 21$, so that the general $f()$ function in Equation (2) here takes the simple form of i.i.d. random intercepts. To complete the model we assign the hyperparameter a vague Gamma prior $\tau_i \sim$ Gamma(a, b) with $a = 1$ and $b = 0.001$.

As explained in Section 2, when specifying the model in the `INLA` library a tiny random noise ϵ_i with zero mean and known high precision is always added to the linear predictor, so the latent Gaussian field for the current example is $\mathbf{x} = \{\eta_1, \ldots, \eta_m, \beta_0, \ldots, \beta_3, u_1, \ldots, u_m\}$, while the vector of hyperparameters has dimension one, $\theta = \{\tau_u\}$.

To run the model using `INLA` two steps have to be taken. Firstly, the linear predictor of the model has to be specified as a `formula` object in `R`. Here the function `f()` is used to specify any possible form of the general $f()$ function in (2). In the current model the i.i.d. random effect is specified using `model="iid"`.

```
>formula = r ~ x1*x2 + f(plate, model="iid")
```

Secondly, the specified model can be run by calling the `inla()` function:

```
>mod.seeds = inla(formula, data=Seeds, family="binomial",
+                 Ntrials=n)
```

The $a = 1$ and $b = 0.001$ parameters for the Gamma prior for τ_u are the default choice, therefore, there is no need to specify them. A different choice of parameters a and b can be specified as `f(plate,model="iid",param=c(a,b))`.

A `summary()` function is available to inspect results:

```
> summary(mod.seeds)

Fixed effects:
                  mean       sd 0.025quant 0.975quant         kld
(Intercept) -0.554623 0.140843 -0.833317 -0.277369 5.27899e-05
x1           0.129510 0.243002 -0.326882  0.600159 3.97357e-07
x2           1.326120 0.199081  0.938824  1.725244 3.98097e-04
x1:x2       -0.789203 0.334384 -1.452430 -0.135351 3.35429e-05

Random effects:
Name       Model          Max KLD
plate    IID model    4e-05

Model hyperparameters:
```

```
                    mean     sd        0.025quant 0.975quant
Precision for plate 1620.89 2175.49  104.63     7682.16

Expected number of effective parameters(std dev): 5.678(2.216)
Number of equivalent replicates : 3.698
```

Standard `summary()` output includes posterior mean, standard deviation, 2.5% and 97.5% quantiles both for the elements in the latent field and for the hyperparameters. Moreover, the expected number of effective parameters, as defined in [21], and the number of data points per expected number of effective parameter (`Number of equivalent replicates`) is also provided. These measures might be useful to assess the accuracy of the approximation, see [20] for more details. Briefly, a low number of equivalent replicates might flag a "difficult" case for the INLA approach.

The `INLA` library includes also a set of functions which post-process the marginal densities obtained by `inla()`. These functions allow computation of the quantiles, percentiles, expectations of function of the original parameter, density of a particular value and also allow sampling from the marginal. As an example consider the following: The output of the `inla()` function provides us with posterior mean and standard deviation of the precision parameter τ_u. Assume that we are instead interested in the posterior mean and standard deviation of the variance parameter $\sigma_u^2 = 1/\tau_u$. This can be easily done by selecting the appropriate posterior marginal from the output of the `inla()` function:

```
prec.marg = mod.seeds$marginals.hyperpar$"Precision for plate"
```

and then using the function `inla.expectation()`

```
> m1 = inla.expectation(function(x) 1/x, prec.marg)
> m2 = inla.expectation(function(x) (1/x)^2, prec.marg)
> sd = sqrt(m2 - m1^2)
>  print(c(mean=m1, sd=sd))
       mean          sd
0.001875261 0.002823392
```

Sampling from posterior densities can be also be done using `inla.rmarginal()`. For example, a sample of size 1000 from the posterior $\widetilde{\pi}(\beta_1|\mathbf{y})$ is obtained as follows:

```
> dens = mod.seeds$marginals.fixed$x1
> sample = inla.rmarginal(1000,dens)
```

More information about functions operating on marginals can be found by typing `?inla.marginal`.

5.2 Childhood undernutrition in Zambia: spatial analysis

In the second example we consider three different spatial models to analyse the Zambia data set presented in [14]. Here the authors study childhood undernutrition in 57 regions of Zambia. A total of $n_d = 4847$ observation are included in the data set. Undernutrition is measured by stunting, or inefficiency height for age, indicating chronic undernutrition. Stunting for child $i = 1, \dots, n_d$ is determined using a Z score defined as

$$Z_i = \frac{AI_i - MAI}{\sigma},$$

where AI refers to the child's anthropometric indicator, MAI refers to the median of the reference population and σ refers to the deviation of the standard population. In addition, the data set includes a set of covariates such as the age of the child (age_i), the body mass index of the child's mother (bmi_i), the district the child lives in (s_i) and four additional categorical covariates. For more details about the data set see [13] and [14].

We assume the scores Z_i (vn: `hazstd`) to be conditionally independent Gaussian random variables with unknown mean η_i and unknown precision τ_z. We consider three different models for the mean parameter η_i. The first is defined as:

$$\eta_i = \mu + \mathbf{z}_i^T \beta + f_s(s_i) + f_u(s_i). \tag{5}$$

This model will be called MOD1. It assumes all six covariates to have a linear effect. Moreover, it contains a spatially unstructured component $f_u(s_i)$ (vn: `distr.unstruct`), which is i.i.d normally distributed with zero mean and unknown precision τ_u, and a spatially structured component $f_s(s_i)$ (vn: `district`) which is assumed to vary smoothly from region to region. To account for such smoothness $f_u(s_i)$ is modeled as an intrinsic Gaussian Markov random field with unknown precision τ_s, see [18]. This specification is also called a conditionally autoregressive (CAR) prior [1] and was introduced by [2]. To ensure identifiability of the mean μ, a sum-to-zero constrain must be imposed on the $f_s(s_i)$'s. The latent Gaussian field for this model is $\mathbf{x} = \{\mu, \{\beta_k\}, \{f_s(\cdot)\}, \{f_u(\cdot)\}, \{\eta_i\}\}$, while the hyperparameters vector is $\theta = \{\tau_z, \tau_u, \tau_s\}$. Vague independent Gamma priors are assigned to each element in θ. When specifying the model in (5) using the `INLA` library, the type of smooth effect is specified using `model="iid"` for the unstructured spatial component and `model="besag"` for the structured one. Moreover, for the spatially structured term, a graph file (e.g. `"zambia.graph"`) containing the neighbourhood structure has to be specified. The structure of such graph file is described in [16]. The resulting model specification looks like:

```
>formula = hazstd ~ edu1 + edu2 + tpr + sex + bmi + agc +
+        f(district, model="besag", graph.file="zambia.graph") +
+        f(distr.unstruct, model="iid")
```

Note that in the `INLA` library a sum-to-zero constraint is the default choice for every intrinsic model. One requirement of the `INLA` library is that, each effect specified through an `f()` function in the formula should correspond to a different column in the data file, that is why the two column `district` and `distr.unstruct` are needed. Fitting the model is done by calling the `inla()` function:

```
> mod = inla(formula, family="gaussian", data=Zambia,
+              control.compute=list(dic=TRUE, cpo=TRUE))
```

The `dic=TRUE` flag makes the `inla()` function compute the model's deviance information criterion (DIC). This is a measure of complexity and fit introduced in [21] and used to compare complex hierarchical models. It is defined as:

$$\mathrm{DIC} = \overline{\mathrm{D}} + p_D,$$

where $\overline{\mathrm{D}}$ is the posterior mean of the deviance of the model and p_D is the effective number of parameters. Smaller values of the DIC indicate a better trade-off between complexity and fit of the model.

The `cpo=TRUE` flag tells the `inla()` function to compute also some predictive measures for the observed y_i given all other observations. In particular the predictive density $\pi(y_i|\mathbf{y}_{-i})$ (called `cpo`) and the probability integral transform $\mathrm{PIT}_i = \mathrm{Prob}(y_i^{\mathrm{new}} < y_i|\mathbf{y}_{-i})$ (called `pit`) are computed. These quantities can be useful to assess the predictive power of the model or to detect surprising observations. See [20] for details on how such quantities are computed. As noted in [12] the simplified Laplace approximation might, in some cases, not be accurate enough when computing predictive measures. The `inla()` function outputs a vector (which can be recovered as `mod$failure`) which contains values from 0 to 1 for each observation. The value 0 indicates that the computation of `cpo` and `pit` for the corresponding observation was computed without problems. A value greater than 0 instead, indicates that there were some computing problems and the predictive measures should be recomputed. See the FAQ section on `www.r-inla.org` for further interest concerning this topic.

The posterior mean for the β parameters, together with standard deviations and quantiles are presented in Table 1. The `inla()` function returns the whole posterior density for such parameters, therefore, if needed other quantities of interest can also be computed. The posterior mean of smooth and unstructured spatial effects are displayed in Figure 1. The output of the `inla()` function also includes posterior marginals for the hyperparameters of the model and posterior marginals for the linear predictor, which are not displayed here. The value of the DIC for MOD1 is displayed in Table 2. To assess the predictive quality of the model the cross-validated logarithmic score [11] can be used. It can be computed using the `inla()` output as:

```
> log.score = -mean(log(mod1$cpo))
```

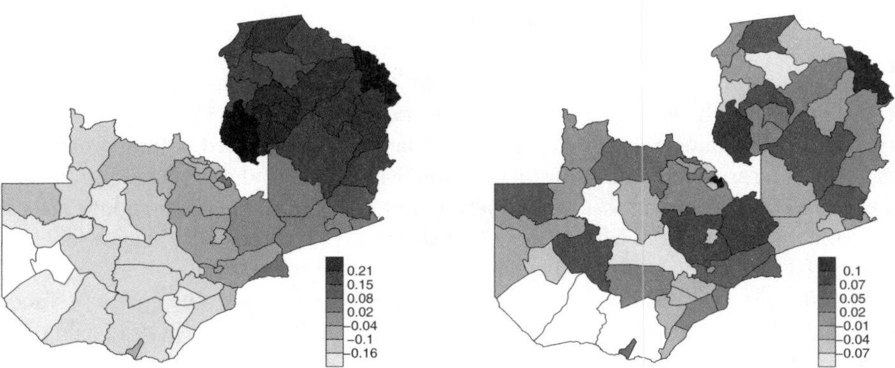

Fig. 1. Posterior mean for the smooth spatial effect (left) and posterior mean for the unstructured spatial effect (right) in MOD1

Table 1. Posterior mean (standard deviation) together with 2.5% and 97.5% quantiles for the linear effect parameters in the three models for the Zambia data

Model	*Covariate*	*Mean(sd)*	*2.5% quant*	*97.5% quant*
	μ	−0.010(0.100)	−0.207	0.187
MOD1	β_{agc}	−0.015(0.001)	−0.017	−0.013
	β_{edu1}	−0.061(0.027)	−0.114	−0.009
	β_{edu2}	0.227(0.047)	0.134	0.320
	β_{tpr}	0.113(0.021)	0.072	0.155
	β_{sex}	−0.059(0.013)	−0.086	−0.033
	β_{bmi}	0.023(0.004)	0.014	0.031
	μ	−0.412(0.096)	−0.602	−0.223
MOD2	β_{edu1}	−0.060(0.026)	−0.111	−0.009
	β_{edu2}	0.234(0.046)	0.145	0.324
	β_{tpr}	0.105(0.021)	0.064	0.145
	β_{sex}	−0.058(0.013)	−0.084	−0.033
	β_{bmi}	0.021(0.004)	0.013	0.029
	μ	−0.366(0.096)	−0.556	−0.178
MOD3	β_{edu1}	−0.061(0.026)	−0.112	−0.010
	β_{edu2}	0.232(0.046)	0.142	0.321
	β_{tpr}	0.107(0.023)	0.062	0.152
	β_{sex}	−0.059(0.013)	−0.084	−0.033

Table 2. DIC value and logarithmic score for the three model in the Zambia example

	MOD1	*MOD2*	*MOD3*
Mean of the deviance	13030.66	12679.90	12672.22
Deviance of the mean	12991.61	12630.89	12610.68
Effective number of parameters	39.04	49.01	61.53
DIC	13069.71	12728.92	12733.76
log Score	1.357	1.313	1.314

The resulting value is displayed in Table 2. A smaller value of the logarithmic score indicates a better prediction quality of the model. The `mod$failure` vector, in this case, contains only 0's so predictive quantities can be used without problems. A tool to assess the calibration of the model is to check the `pit` histogram. As suggested in [6], in fact, in a well calibrated model, the `pit` values should have a uniform distribution. For MOD1 the `pit` histogram (not shown here but available on `www.r-inla.org`) doesn't show any sign of wrong calibration.

As discussed in Section 3 the default integration strategy in the `inla()` function is the CCD strategy. It is possible to choose a GRID strategy instead using the following call to `inla()`:

```
> mod = inla(formula, family="gaussian", data=Zambia,
+              control.inla = list(int.strategy = "grid"))
```

The computational time increases from ca 9 seconds needed by the CCD integration to ca 16 seconds, while a comparison of the results coming from the two fits (not shown here) does not present any significant difference.

As discussed in [14] there are strong reasons to believe that the effect of the age of the child is smooth but not linear. To check such an assumption we can modify the previous model to:

$$\eta_i = \mu + \mathbf{z}_i^T \beta + f_1(\text{age}_i) + f_s(s_i) + f_u(s_i). \tag{6}$$

This extended model will be called MOD2. Here $\{f_1(\cdot)\}$ follows an intrinsic second-order random-walk model with unknown precision τ_1, see [18]. To ensure identifiability of μ, a sum-to-zero constraint must be imposed on $f_1(\cdot)$. The latent field is then $\mathbf{x} = \{\mu, \{\beta_k\}, \{f_s(\cdot)\}, \{f_u(\cdot)\}, \{f_1(\cdot)\}, \{\eta_i\}\}$ while the hyperparameter vector is $\theta = \{\tau_z, \tau_u, \tau_s, \tau_1\}$. The INLA specification of MOD2 differs form the previous one simply because now an f() function is used also to define the smooth effect of age.

```
>formula = hazstd ~ edu1 + edu2 + tpr + sex + bmi +
+       f(agc, model="rw2") +
+       f(district, model="besag", graph.file="zambia.graph") +
+       f(distr.unstruct, model="iid")
```

The call to the inla() function is not changed. The estimated posterior mean and quantiles of the non-linear effect of age is plotted in Figure 2(a) and the non-linearity of the age effect is clear. The improvement obtained by using a more flexible model for the effect of the age covariate can be seen also from the decreased value of the DIC in Table 2. This second model results also to be more powerful as a prediction tool, as indicated by the decreased value of the logarithmic score in Table 2. Also for this model the predictive quantities are computed without problems and the pit distribution is close to uniform. The estimates of the other parameters in the model are reported in Table 1, the estimated spatial effects are similar to that in MOD1 and

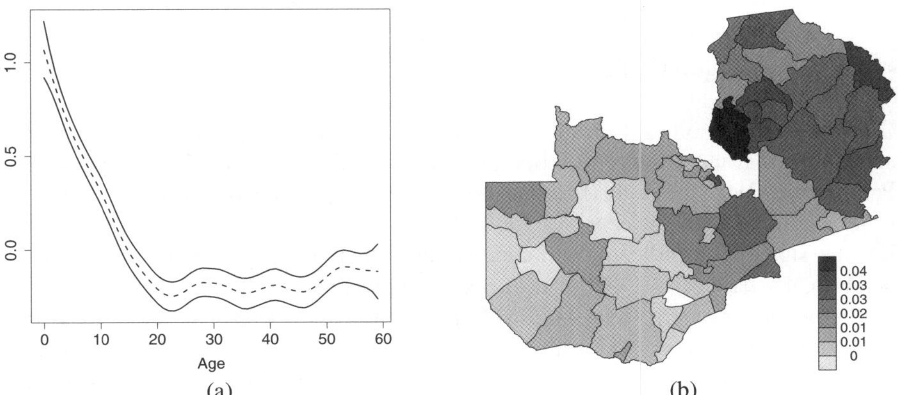

(a) (b)

Fig. 2. (a) Estimated effect of age (posterior mean together with 2.5% and 97.5% quantiles) using MOD2; **(b)** Posterior mean for the space-varying regression coefficient in MOD3

are not reported here. Results are similar to those obtained using BayesX, a program which performs a MCMC study [13].

As an alternative hypothesis to explain spatial variability one could imagine that the effect of one covariate, for example the mother's bmi, although being linear has a different slope for different regions. Again we assume this spatial variability to be smooth. These kinds of models are known as space-varying regression models [10]. Using the notation in equation (2) the model (MOD3) can be written as:

$$\eta_i = \mu + \mathbf{z}_i^T \beta + f_1(\text{age}_i) + \text{bmi}_i \; f_2(s_i). \tag{7}$$

We assume here that the whole spatial variability is explained by the space varying regression parameter so that no other spatial effect is needed. Moreover, we assume the age covariate to have a non-linear effect. The model for $f_1(\cdot)$ is as in MOD2 while for $f_2(\cdot)$ we assume a `"besag"` model, this time the sum-to-zero constraint is not necessary since there are no identifiability problems. Here the `bmi` covariate simply acts as a known weight for the IGMRF $f_2(\cdot)$. The `INLA` specification of the model is as follows:

```
>formula = hazstd ~ edu1 + edu2 + tpr + sex +
+         f(agc, model="rw2") +
+         f(district, bmi, model="besag",
+         graph.file="zambia.graph",
+         constr=FALSE)
```

The order of `district` and `bmi` in the second of the `f()` functions of the formula above is important since arguments are matched by position: the first argument is always the latent field and the second is always the weights. Note, moreover, that the sum-to-zero constraint has to be explicitly removed since, as said before, is default for all intrinsic models. The resulting estimates for the space varying regression parameter are displayed in Figure 2(b).

The computing time (using the default CCD strategy) goes from a minimum of 9 seconds for MOD1 to a maximum of 14 seconds for MOD2.

5.3 A simple example of survival data analysis

Our last example comes from survival analysis literature. A typical setting in survival analysis is that we observe the time point t at which the death of a patient occurs. Patients may leave the study (for some reason) before they die. In this case the survival time is said to be right censored, and t refers to the time point when the patient left the study. The indicator variable δ is used to indicate whether t refers to the death of the patient ($\delta = 1$) or to a censoring event ($\delta = 0$). The key quantity in modeling the probability distribution of t is the hazard function $h(t)$, which measures the instantaneous death rate at time t. We also define the cumulative hazard function $H(t) = \int_0^t h(s)\,ds$, implicitly assuming that the study started at time $t = 0$. A different starting time can also be considered and it is usually referred to as truncation time. The log-likelihood contribution from one patient is $\delta \log(h(t)) - H(t)$. A commonly used model for $h(t)$ is Cox's proportional hazard model [4], in which the hazard rate for the ith patient is assumed to be in the form

$$h_i(t) = h_0(t)\exp(\eta_i), \quad i = 1, \dots, n.$$

Here, $h_0(t)$ is the "baseline" hazard function (common to all patients) and η_i is a linear predictor. In this example we shall assume that the baseline hazard belongs to the Weibull family: $h_0(t) = \alpha t^{\alpha-1}$ for $\alpha > 0$.

In [17] this model is used to analyse a data set on times to kidney infection for a set of $n_d = 38$ patients. The data set contains two observations per patient (the time to first and second recurrence of infection). In addition there are three covariates: "age" (continuous), "sex" (dichotomous) and "type of disease" (categorical, four levels), and an individual specific random effect (vn: ID), often named frailty: $u_i \sim N(0, \tau^{-1})$. Thus, the linear predictor becomes

$$\eta_i = \beta_0 + \beta_{sex}\text{sex}_i + \beta_{age}\text{age}_i + \beta_D \mathbf{x}_i + u_i, \tag{8}$$

where $\beta_D = (\beta_2, \beta_3, \beta_4)$ and $\mathbf{x}_i$ is a dummy vector coding for the disease type. Here we used a corner-point constraint imposing $\beta_1 = 0$.

Fitting a survival model using INLA is done using the following commands:

```
>formula = inla.surv(time,event) ~ age + sex + dis2 + dis3 +
+            dis4 + f(ID, model="iid")
>mod = inla(formula, family="weibull", data=Kidney)
```

Note that the function inla.surv() is needed to define the response variable of a survival model. This function is used to define different censoring schemes such as right, left or interval censoring plus, possibly, truncation times. Including more complex effects in model (8) such as, for example, smooth effects of covariates or spatial effects can be done in exactly the same way as for the previous examples. Posterior means and standard deviations, together with quantiles, for the model parameters are shown in Table 3 and are similar to those obtained by Gibbs sampling via WinBUGS and by maximum likelihood. Fitting the model took less than 2 seconds.

Table 3. Posterior mean, standard deviation and quantiles for the parameters in the survival data example

	mean	*sd*	*2.5% quant*	*97.5% quant*
β_0	−4.809	0.954	−6.77	−3.103
β_{age}	0.003	0.016	−0.028	0.036
β_{sex}	−2.071	0.535	−3.180	−1.076
β_2	0.155	0.591	−1.007	1.346
β_3	0.679	0.595	−0.467	1.900
β_4	−1.096	0.863	−2.813	0.611
α	1.243	0.151	0.970	1.560
τ	2.365	1.647	0.586	6.976

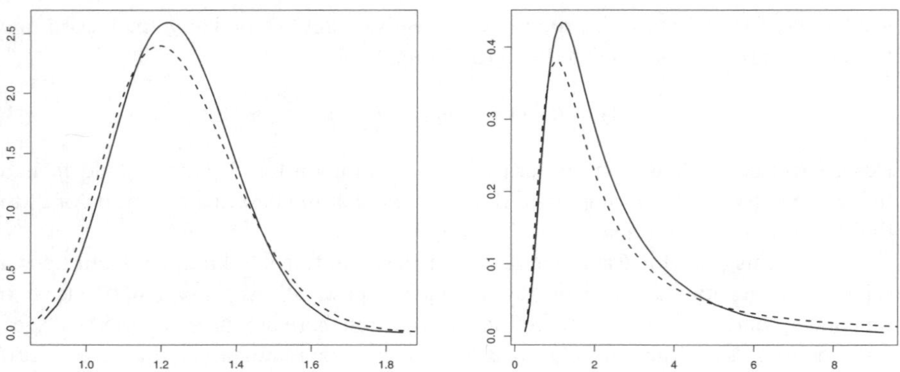

Fig. 3. Estimate of the posterior marginals for θ basic estimation (*solid line*) and improved one (*dashed line*). Left: posterior marginals for α. Right: posterior marginal for τ

The `inla()` function computes the posterior marginals for the hyperparameters $\theta = (\alpha, \tau)$ using only a few points in the θ space (CCD strategy). As noted in Section 3 this might be not accurate enough. The function `inla.hyperpar()`, which takes as input the output of `inla()`, recomputes the marginals for the hyperparameters in a more accurate way, using a grid integration:

```
> hyperpar = inla.hyperpar(mod)
```

In cases where the dimension m of θ is large, computing posterior marginals using `inla.hyperpar()` can be time consuming. In this case, being $m = 2$, recomputing marginals for θ took less than 3 seconds. Figure 3 shows the two approximations for the hyperparameters in model 8. For the current example there is no particular improvement from using the more accurate approximation computed by `inla.hyperpar()`.

The `INLA` library can also deal with exponential models for the baseline function $h_0(t)$. Semiparametric models for $h_0(t)$ such as the piecewise log-constant are, at the moment, under study.

6 Conclusions

As shown in this paper INLA is a powerful inferential tool for latent Gaussian models. The computational core of the `inla` programme treats any kind of latent field in the same way thus behaving as a black-box. The available `R` interface `INLA`, can easily be handled by the user to obtain fast and reliable estimates. The large series of different options both for the approximations of the posterior marginals of the latent field, and for the exploration of the hyperparameter space may generate confusion in the novice user. On the other hand, the default choices in the `INLA` library, usually offer a good starting point for the analysis.

The `INLA` library computes also quantities useful for model comparison, a feature that becomes important when the computational speed gives the possibility to fit several models to the same data set.

The `INLA` library contains also functions to process the posterior marginals obtained by the `inla()` function, so that it is possible to compute quantiles, percentiles or expectations of functions of the original random variable. Sampling from such posterior marginals is also possible.

References

1. Banerjee, S., Carlin, B., Gelfand, A.: Hierarchical Modeling and Analysis for Spatial Data. Chapman & Hall/CRC, Boca Raton, Florida, USA (2004)
2. Besag, J., York, J., Mollie, A.: Bayesian image restoration with two application in spatial statistics. Annals of the Institute in Mathematical Statistics **43**(1), 1–59 (1991)
3. Breslow, N.E., Clayton D.G.: Approximate inference for Generalized Linear Mixed models. Journal of the American Statistical Association **88**, 9–25 (1993)
4. Cox, D.R.: Regression models and life-tables. Journal of the Royal Statistical Society, Series B, **34**, 187–220 (1972)
5. Crowder, M.J.: Beta-Binomial ANOVA for Proportions. Applied Statistics **27**, 34–37 (1978)
6. Czado, C., Gneiting, T., Held, L.: Predictive model assessment for count data. Biometrics **65**, 1254–1261 (2009)
7. Eidsvik, J., Martino, S., Rue, H.: Approximate Bayesian Inference in spatial generalized linerar mixed models. Scandinavian Journal of Statistics **36**, (2009)
8. Fahrmeir, L., Tutz, G.: Multivariate Statistical Modeling Based on Generalized Linear Models. Springer-Verlag, Berlin (2001)
9. Fong, Y., Rue H., Wakefield J.: Bayesian inference for Generalized Linear Mixed Models. Biostatistics, doi:10.1093/biostatistics/kxp053 (2009)
10. Gamerman, D., Moreira, A., Rue, H.: Space-varying regression models: specifications and simulation. Computational Statistics & Data Analysis **42**, 513–533 (2003)
11. Gneiting, T., Raftery, A.E.: Strictly proper scoring rules, prediction and estimation. Journal of American Statistical Association, Series B, **102**, 359–378 (2007)
12. Held, L., Schrodle, B., Rue, H.: Posterior and cross-validatory predictive checks: A comparison of MCMC and INLA. In: Kneib, T., Tuts, G. (eds) *Statistical Modeling and Regression structures – Festschrift in Honour of Ludwig Fahrmeir*. Physica-Verlag, Heidelberg (2010)
13. Kneib, T., Lang, S., Brezger, A.: Bayesian semiparametric regression based on MCMC techniques: A tutorial (2004) http://www.statistik.uni-muenchen.de/~lang/Publications/bayesx/mcmctutorial.pdf
14. Kandala, N.B., Lang, S., Klasen, S., Fahrmeir, L.: Semiparametric Analysis of the Socio-Demographic and Spatial Determinants of Undernutrition in Two African Countries. Research in Official Statistics **1**, 81–100 (2001)
15. Lunn, D.J., Thomas, A., Best, N., Spiegelhalter, D.: WinBUGS – a Bayesian modelling framework: concepts, structure, and extensibility. Statistics and Computing **10**, 325–337 (2000)

16. Martino, S., Rue, H.: Implementing Approximate Bayesian inference using Integrated Nested Laplace Approximation: A manual for the INLA program. Technical report, Norwegian University of Science and Technology, Trondheim (2008)
17. McGilchrist, C.A., Aisbett, C.W.: Regression with frailty in survival analysis. Biometrics **47**(2), 461–466 (1991)
18. Rue, H., Held, L.: Gaussian Markov Random Fields: Theory and Applications. Vol. **104** of Monographs on Statistics and Applied Probability, Chapman&Hall, London (2005)
19. Rue, H., Martino, S.: Approximate Bayesian Inference for Hierarchical Gaussian Markov Random Fields Models. Journal of Statistical Planning and Inference **137**, 3177–3192 (2007)
20. Rue, H., Martino, S., Chopin, N.: Approximate Bayesian Inference for Latent Gaussian Models Using Integrated Nested Laplace Approximations (with discussion). Journal of the Royal Statistical Society, Series B, **71**, 319–392 (2009)
21. Spiegelhalter, D., Best, N., Bradley, P., van der Linde, A.: Bayesian measure of model complexity and fit (with discussion). Journal of the Royal Statistical Society, Series B, **64**(4), 583–639 (2002)
22. Tierney, L., Kadane, J.B.: Accurate approximations for posterior moments and marginal densities. J. Am. Statist. Ass. **81**, 82–86 (1986)

A graphical models approach for comparing gene sets

M. Sofia Massa, Monica Chiogna and Chiara Romualdi

Abstract. Recently, a great effort in microarray data analysis has been directed towards the study of the so-called gene sets. A gene set is defined by genes that are, somehow, functionally related. For example, genes appearing in a known biological pathway naturally define a gene set. Gene sets are usually identified from a priori biological knowledge. Nowadays, many bioinformatics resources store such kind of knowledge (see, for example, the Kyoto Encyclopedia of Genes and Genomes, among others). In this paper we exploit a multivariate approach, based on graphical models, to deal with gene sets defined by pathways. Given a sample of microarray data corresponding to two experimental conditions and a pathway linking some of the genes, we investigate whether the strength of the relations induced by the functional links change among the two experimental conditions.

Key words: Gaussian graphical models, gene sets, microarray, pathway

1 Introduction

Microarray technology permits the simultaneous quantification of the expression of thousands of genes in a single experiment. Since the advent of this technology, the primary interest has been directed towards the identification of differentially expressed genes.

Many statistical tests, centred on the null hypothesis of equal expression of a gene between two (or more) experimental conditions, have been proposed in past years; see for example [11] for an extensive review. On sets of genes, the so-called significance analyses typically assess the level of significance for a gene at a time, producing then a list of differentially expressed genes by using a cutoff threshold on the levels of significance. This list is then investigated from a biological point of view, to assess the enrichment of specific biological themes in the list [8]. This is achieved through biologically defined gene sets derived from Gene Ontology (available at `http://www.geneontology.org`) or by means of some pathway databases. Many authors pointed out a series of drawbacks of this approach. A major drawback is related to the use of a threshold for the identification of differentially expressed genes, and, therefore, gene sets. For example, [9] show that different choices of threshold

Mantovan, P., Secchi, P. (Eds.): Complex Data Modeling and Computationally Intensive Statistical Methods

highly affect the biological conclusions of the study. Secondly, the use of a strict cutoff value can limit the output of the analysis. [3] show that the strict cutoff value does not permit consideration of many genes with moderate, but meaningful expression changes, and this reduces the statistical power for the identification of true positives. Finally, such approaches are all based on the incorrect assumption of independent gene sampling, which, of course, increases false positive predictions.

In recent years, interest has moved from the study of individual genes to that of groups of genes (for example, pathways) and methods for gene set analysis have received great attention. The aim, in this case, is to identify groups of genes with moderate, but coordinated expression changes, which should enable the understanding of cellular processes such as regulatory networks of functionally related components. Such approaches directly score pre-defined gene sets for differential expression, and, therefore, are free from the problems of the cutoff-based methods. Several gene set analysis methods have recently been developed both in the univariate and multivariate context; see [1], and references therein, for a comprehensive review on existing methods. Usually, two kinds of null hypothesis exist for testing the coordinated association of gene sets with a phenotype of interest. The first type, called *competitive*, assumes the same level of association of a gene set with the given phenotype as the complement of the gene set. The second type, called *self-contained*, considers only the genes within a gene set and assumes that there is no gene in the gene set associated with the phenotype.

In this study, we will focus on gene set analysis within a multivariate approach based on graphical models. The graphical structure of a pathway will be used to develop a test to compare two experimental conditions of interest. The paper is organised as follows. In Section 2 we give a brief introduction to the structure of pathways. Section 3 presents the data and our graphical models approach in detail. Section 4 proposes a test in this specific context, and Section 5 gives some results and final remarks.

2 A brief introduction to pathways

Even if there is not a precise definition, a biological pathway can be described as a set of linked biological components interacting with each other over time to generate a single biological effect. Participants in one pathway can be involved also in others, leading to dependent pathways. The Kyoto Encyclopedia of Genes and Genomes (KEGG, available at `http://www.genome.jp/kegg/`, [5]) is one of the most widely used pathways database, with more than a hundred pathways and more than fifty available signaling pathways. Figure 1 represents one of such pathways, the B Cell receptor signaling pathway. As one can see, it is composed by edges and nodes, which have the following meanings. Usually, rectangles are gene products, mostly proteins, but including RNA. For example, the rectangle referring to CaN contains the genes: CHP, PPP3CA, PPP3CB, PPP3CC, PPP3R1, CHP2. The edges between rectangles are protein-protein or protein/RNA interactions and they can be:

- undirected, to represent binding or association;
- directed, to represent activation;

- directed with a +p, to represent phosporylation;
- directed with a −p, to represent dephosporylation;
- dashed and directed, to represent indirect effects;
- directed with a bar at the end, to represent inhibition.

Circles are other types of molecules, mostly chemical compounds, while the big white rectangles represent the link to other pathways.

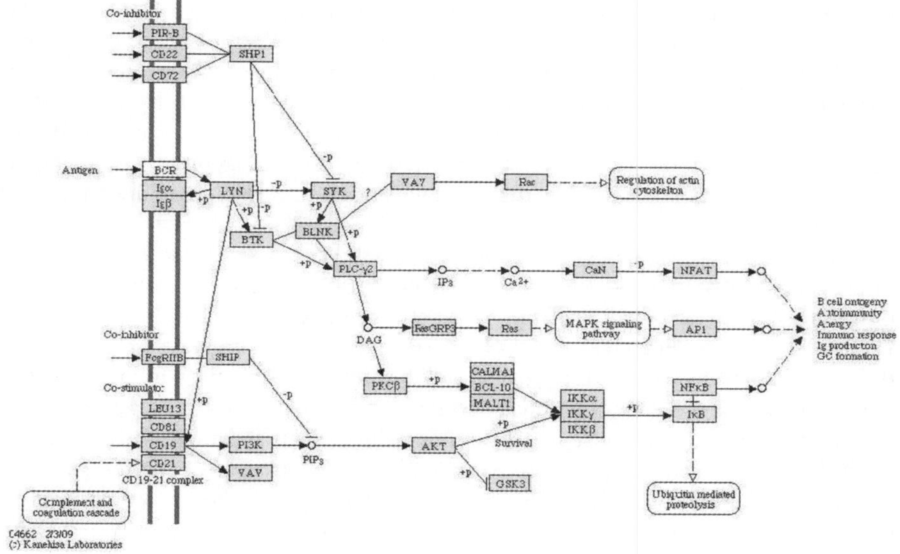

Fig. 1. B Cell receptor signaling pathway taken from the Kyoto Encyclopedia of Genes and Genomes (KEGG) [5]

The increasing number of scientific results on specific aspects of cellular processes makes pathway topology a dynamic entity. A pathway should be updated monthly on the basis of the most recent scientific findings. Although KEGG database represents a valid and exhaustive repository of biological pathways for several organisms, sometimes pathway topologies are largely simplified and rarely updated. For this reason, collaborative projects were born recently in order to create groups of experts that manually care for each single pathway. One of the results of these collaborative projects is the creation of an open platform, called WikiPathways [10], where pathways are extremely detailed and can also be edited. The use of both KEGG and WikiPathways information (available at `http://www.wikipathways.org`) provides a good compromise between graph simplicity and structure accuracy.

3 Data and graphical models setup

In this study, we use a dataset recently published by [4], which characterises gene expression signatures in acute lymphocytic leukemia (ALL) cells associated with known genotypic abnormalities in adult patients. Several distinct genetic mechanisms lead to acute lymphocytic leukemia (ALL) malignant transformations deriving from distinct lymphoid precursor cells that have been committed to either T-lineage or B-lineage differentiation. Chromosome translocations and molecular rearrangements are common events in B-lineage ALL and reflect distinct mechanisms of transformation. The relative frequencies of specific molecular rearrangements differ in children and adults with B-lineage ALL. The B Cell Receptor (BCR/ABL) gene rearrangement occurs in about 25% of cases in adult ALL, and much less frequently in pediatric ALL. Because these cytogenetic abnormalities reflect distinct mechanisms of transformation, molecular differences between these two types of rearrangements could help to explain why children and adults with ALL have such different outcomes following conventional therapy.

Data are freely available at the Bioconductor website. Expression values, appropriately normalised and filtered, derived from Affymetrix single channel technology, consist of 37 observations from one experimental condition (BCR; presence of BCR/ABL gene rearrangement) and 41 observations from another experimental condition (NEG; absence of rearrangement) and the aim is to study the two experimental conditions for the presence or absence of rearrangement. As we can see, the gene BCR is central in this study, because it is the only one involved in the process of rearrangement. For this reason, we decide to focus on the B cell receptor signaling pathway (represented in Figure 1) which has the gene BCR as input. If there is a modification of this gene, we expect that also the genes belonging to the connected pathway will be highly influenced from the BCR/ABL gene rearrangement. For simplicity, instead of considering the whole pathway, we choose a subset of it containing only 13 gene products. In the following, we will refer to the chosen subset as to the pathway of interest.

The study of the behaviour of the pathway in the two experimental conditions will be pursued in a graphical models context. We believe that this approach, which is still largely unexplored, goes in a direction which can valuably complement approaches more extensively offered by the current literature. We are not interested in detecting the structure of the pathway, because we consider it as fixed from the very beginning. Our aim is to use a statistical test to compare the strength of the links of the pathway in the two experimental conditions. Therefore, we assume also that the structure of the pathway does not change between the two conditions.

To begin with, we convert the structure of the pathway into a simple directed acyclic graph (DAG), by following these simple steps: if in one rectangle (complex) there are multiple genes, we consider only one of them as representative; since the edges with a bar at the end represent inhibition, we interpret them as arrows, and the arrows with +p and −p are considered as simple arrows. Then, we use WikiPathways to derive the orientation of all undirected edges. With this extra information, it is always possible to convert pathways into DAGs, checking always for the absence of direct cycles.

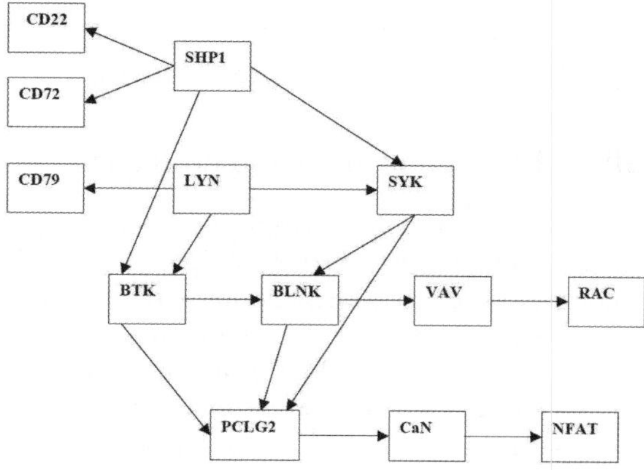

Fig. 2. DAG D corresponding to 13 proteins of the B cell receptor signaling pathway

The DAG D corresponding to the B cell receptor signaling pathway is shown in Figure 2. Starting from D, we derive its moral graph D^m (Figure 3). A moral graph can differ from the corresponding DAG, in the sense that it usually has more edges and that they are all undirected, but the choice of working with a moral graph does not affect the purpose of this study. From now on, we base our analyses on the undirected graph D^m that we indicate as G. We assume to model the data from the two experimental conditions (BCR and NEG) with two graphical Gaussian models ([6]) with the same undirected graph G,

$$\mathcal{M}_1(G) = \{Y \sim N_{13}(0, \Sigma_1),\ \Sigma_1^{-1} \in S^+(G)\} \quad \text{and}$$
$$\mathcal{M}_2(G) = \{Y \sim N_{13}(0, \Sigma_2),\ \Sigma_2^{-1} \in S^+(G)\},$$

respectively, where $S^+(G)$ is the set of symmetric positive definite matrices with null elements corresponding to the missing edges of G.

As we said before, we are interested in comparing the strength of the links in the two experimental conditions. In a graphical Gaussian models context this is simply achieved by comparing the two concentration matrices (inverse of the covariance matrices), because they contain all the information about the underlying structure.

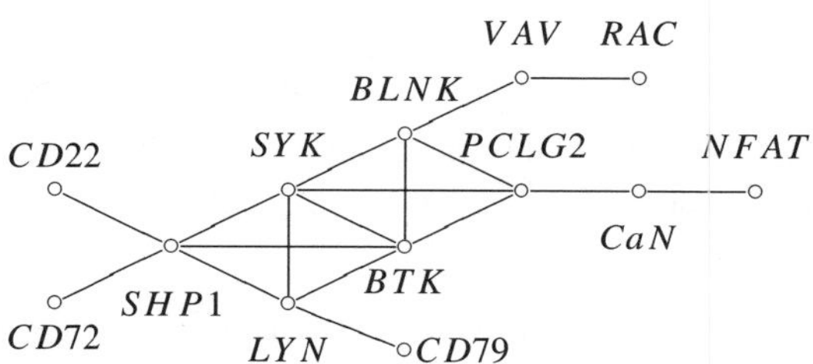

Fig. 3. Moral graph D^m corresponding to 13 proteins of the B cell receptor signaling pathway

Therefore, the interest is in testing the hypothesis $\Sigma_1^{-1} = \Sigma_2^{-1}$ and this will be the topic of the next section.

4 Test of equality of two concentration matrices

Assuming that gene expression data on two experimental conditions come from two multivariate normal distributions (relative or absolute gene expression measurements are approximately normal on the log scale) according to the common undirected graph G, we aim to verify the equality of the concentration matrices of the two graphical Gaussian models. The following methodology is the transposition of the methods for comparing covariance matrices [2] in the specific case of graphical Gaussian models.

Suppose we have $y_1 = (y_1^j)$, $j = 1, \cdots, n_1$ observations from $N_p(0, \Sigma_1)$ and $y_2 = (y_2^j)$, $j = 1, \cdots, n_2$ observations from $N_p(0, \Sigma_2)$, $\Sigma_1^{-1} = K_1$ and $\Sigma_2^{-1} = K_2$ with $K_1, K_2 \in S^+(G)$.
We want to test the hypothesis

$$H_0 : K_1 = K_2 \quad \text{against} \quad H_1 : K_1 \neq K_2.$$

If we set $W_i = \sum_{j=1}^{n_i} (y_i^j)(y_i^j)^T$, $i = 1, 2$, the likelihood function, $L(K_1, K_2)$, is

$$L(K_1, K_2) = \prod_{i=1}^{2} (2\pi)^{-\frac{n_i p}{2}} (\det K_i)^{\frac{n_i}{2}} e^{-\frac{1}{2}\mathrm{tr}(K_i W_i)},$$

and each block of it may be maximised separately [2]. In more detail, the estimates $\hat{K}_1$ and $\hat{K}_2$ are computed by direct calculation (if the graph is decomposable) or, in general, by using the Iterative Proportional Scaling algorithm [6]. Under the alternative hypothesis, the algorithm uses the sample covariance matrices

$$S_1 = (n_1 - 1)^{-1} \cdot W_1 \quad \text{and} \quad S_2 = (n_2 - 1)^{-1} \cdot W_2,$$

and computes $\hat{\Sigma}_1$ and $\hat{\Sigma}_2$. Under the null hypothesis, it uses the pooled covariance matrix

$$S = (n_1 + n_2 - 2)^{-1}\{(n_1 - 1)S_1 + (n_2 - 1)S_2\},$$

and computes only $\hat{\Sigma}$ that is the common value of K_1 and K_2.

Furthermore $\hat{K}_1 = (\hat{\Sigma}_1)^{-1}$, $\hat{K}_2 = (\hat{\Sigma}_2)^{-1}$, $\hat{K} = (\hat{\Sigma})^{-1}$. The likelihood ratio test, Λ, is

$$\Lambda = \frac{L_{H_0}(\hat{K}_1, \hat{K}_2)}{L_{H_1}(\hat{K}_1, \hat{K}_2)} = \frac{L_{H_0}(\hat{K})}{L_{H_1}(\hat{K}_1, \hat{K}_2)}.$$

If we let $W = W_1 + W_2$, and exploit the fact that $\mathrm{tr}(\hat{K}_i W_i) = n_i \mathrm{tr}(\hat{K}_i \hat{K}_i^{-1}) = n_i p$ and $\mathrm{tr}(\hat{K} W) = (n_1 + n_2)\mathrm{tr}(\hat{K}\hat{K}^{-1}) = (n_1 + n_2)p$ [6], we have

$$\Lambda = \prod_{i=1}^{2} \left(\frac{\det \hat{K}}{\det \hat{K}_i} \right)^{\frac{n_i}{2}}, \quad \text{and} \quad -2\log\Lambda = \sum_{i=1}^{2} n_i \log\left(\frac{\det \hat{K}_i}{\det \hat{K}} \right).$$

The asymptotic distribution of $-2\log\Lambda$ is χ^2_{r+p} where r is the number of edges of G.

5 Conclusions

If we perform the above test with the data described in Section 3 with $n_1 = 37$, $n_2 = 41$, $p = 13$, opportunely scaled to have a mean equal to zero, the null hypothesis is rejected. Indeed, this result confirms that patients with BCR rearrangement show a significant deregulation of the entire pathway centred on the BCR gene products with respect to those without rearrangement.

It is also of interest to check where the two concentration matrices are different. Since the graph is decomposable, we decompose it into its cliques and repeat the previous test for each clique. Note that, in this way, we compare the cliques only marginally. Since the cliques are complete subgraphs, we can directly test the equality of the covariance matrices on the cliques and therefore do not need the test of Section 4. One can directly follow the well-known theory for comparing covariance matrices [2]. Also, the nine tests to be performed are corrected according to Bonferroni's method. The cliques that behave differently between the two experimental conditions are (CD22, CD72), (PCLG2, BLNK), (CaN, NFAT), (SHP1, LYN, SYK, BTK), (SYK, BTK, BLNK, PCLG2).

These results suggest that signaling pathway deregulation starts at CD22 and ends at NFAT, thus underlying the importance of the way out of the pathway given by (CaN, NFAT), that is to say the calcineurin/NFAT signaling pathway. Deregulation of calcineurin/NFAT signaling and/or abnormal expression of its components have recently been reported in solid tumours of epithelial origin, lymphoma and lymphoid leukemia [7], thus validating our approach.

Acknowledgement. This work was funded by the University of Padova under grants CPDR070805 and CPDR075919.

References

1. Ackermann, M., Strimmer, K.: A general modular framework for gene set enrichment analysis. BMC Bioinformatics **10**, 47 (2009)
2. Anderson, T. W.: An introduction to multivariate statistical analysis. Wiley, New York (2003)
3. Ben-Shaul, Y., Bergman, H., Soreq, H.: Identifying subtle interrelated changes in functional gene categories using continuous measures of gene expression. Bioinformatics **21**, 1129–1137 (2005)
4. Chiaretti, S., Li, X., Gentleman, R., Vitale, A., Wang, K. S., Mandelli, F., Fo, R., Ritz, J.: Gene expression profiles of B-lineage adult acute lymphocytic leukemia reveal genetic patterns that identify lineage derivation and distinct mechanisms of transformation. Clinical Cancer Research **11**, 7209–7219 (2005)
5. Kanehisa, M., Goto, S.: KEGG: Kyoto Encyclopedia of Genes and Genomes. Nucleic Acids Research **28**, 27–30 (2000)
6. Lauritzen, S. L.: Graphical models. Clarendon Press, Oxford (1996)
7. Medyouf, H., Ghysdael, J.: The calcineurin/NFAT signaling pathway: a novel therapeutic target in leukemia and solid tumors. Cell Cycle **7**, 297–303 (2009)

8. Nam, D., Kim, S. Y.: Gene-set approach for expression pattern analysis. Briefings in Bioinformatics **9**, 189–197 (2008)
9. Pan, K.H., Lih, C.J., Cohen, S.N.: Effects of threshold choice on biological conclusions reached during analysis of gene expression by DNA microarrays. *Proceedings of the National Academy of Sciences of the United States of America* **102**, 8961–8965 (2005)
10. Pico, A.R., Kelder, T., van Iersel, M.P., Hanspers, K., Conklin, B.R., Evelo, C.: Wikipathways: pathway editing for the people. PLOS Biology **6**, 184 (2008)
11. Shaik, J.S., Yeasin, M.: A unified framework for finding differentially expressed genes from microarray experiments. BMC Bioinformatics **18**, 347 (2007)

Predictive densities and prediction limits based on predictive likelihoods

Paolo Vidoni

Abstract. The notion of predictive likelihood stems from the fact that in the prediction problem there are two unknown quantities to deal with: the future observation and the model parameter. Since, according to the likelihood principle, all the evidence is contained in the joint likelihood function, a predictive likelihood for the future observation is obtained by eliminating the nuisance quantity, namely the unknown model parameter. This paper focuses on the profile predictive likelihood and on some modified versions obtained by mimicking the solutions proposed to improve the profile (parametric) likelihood. These predictive likelihoods are evaluated by studying how well they generate prediction intervals. In particular, we find that, at least in some specific applications, these solution usually improve on those ones based on the plug-in procedure. However, the associated predictive densities and prediction limits do not correspond to the optimal frequentist solutions already described in the literature.

Key words: asymptotic expansion, likelihood, prediction interval, predictive distribution

1 Introduction

This paper investigates properties of prediction procedures based on likelihood. In particular, the goodness of a predictive likelihood is studied by considering the associated predictive densities and prediction intervals, which are evaluated from the frequentist perspective.

Although a preliminary idea of predictive likelihood appears in a paper by R.A. Fisher, the first two contributions on this concept are [12] and [11], where the term predictive likelihood is explicitly introduced. Other papers include [7–9, 14–16]. Indeed, [5] provides a valuable and complete review on a number of predictive likelihoods, while [6] presents philosophical reasons for employing predictive likelihoods, in accordance with a general version of the likelihood principle. This paper focuses on the profile predictive likelihood and some modified versions obtained by approximating a Bayesian predictive density or by mimicking the solutions proposed to improve the profile (parametric) likelihood.

The predictive likelihoods are evaluated by studying how well they generate predictive densities and prediction intervals. More precisely, we normalise the predictive

Mantovan, P., Secchi, P. (Eds.): Complex Data Modeling and Computationally Intensive Statistical Methods

likelihoods to be predictive densities and we obtain an explicit expression for the associated quantiles, which define the so-called prediction limits. These predictive densities and prediction limits correspond to modifications of those obtained using the plug-in procedure.

Evaluation of predictive likelihoods on the grounds of frequentist coverage probabilities of the associated prediction limits, emphasises that these solutions do not necessarily improve on those of the plug-in procedure, as far as the order of the coverage error is considered. In [10] an analogous result is obtained with regard to a particular predictive likelihood, similar to that proposed in [9]. In spite of these theoretical findings concerning high order coverage accuracy, we find that, at least in some specific applications, these solutions usually improve on those based on the plug-in procedure. This fact is also confirmed by some theoretical and empirical results related to prediction within the Gaussian model.

The paper is organises as follows. Section 2 provides a quick review on the plug-in and improved (namely with coverage error reduced to the required order) prediction intervals and on the above mentioned solutions based on predictive likelihood methods. Section 3 presents predictive densities and prediction limits related to these predictive likelihoods, with the associated coverage probabilities. Finally, Section 4 gives examples, with simulation experiments, involving various models related to the Gaussian distribution.

2 Review on predictive methods

Prediction is studied from the frequentist viewpoint and the aim here is to define prediction intervals with coverage probability close to the target nominal value. In particular, we will consider prediction limits, that is quantiles from a predictive distribution, obtained using both the plug-in procedure, and related improvements, and the notion of predictive likelihood.

Let (Y, Z) be a continuous random vector following a joint density $p(y, z; \theta)$, with $\theta \in \Theta$ an unknown d-dimensional parameter; $Y = (Y_1, \dots, Y_n)$ is observable, while Z denotes a future, or yet unobserved, random variable. For ease of exposition we consider $Y = (Y_1, \dots, Y_n)$ as a sample of independent, identically distributed, random variables each with density $f(\cdot; \theta)$ and Z is treated as an independent future random variable with density $g(\cdot; \theta)$, possibly different from $f(\cdot; \theta)$. The distribution function of Z is $G(\cdot; \theta)$. We also assume that $f(\cdot; \theta)$ and $g(\cdot; \theta)$ are sufficiently smooth functions of the parameter θ. An α-prediction interval for Z or, in particular, an α-prediction limit $c_\alpha(y)$ is such that, exactly or approximately,

$$P_{Y,Z}\{Z \leq c_\alpha(Y); \theta\} = \alpha, \tag{1}$$

for all θ, where $\alpha \in (0, 1)$ is fixed. The above probability, called coverage probability, refers to the joint distribution of (Y, Z). If there exists an exact or approximate ancillary statistic, it could be reasonable, according to the conditionality principle for inference on θ, to consider the coverage probability conditional on the ancillary. For the conditional approach see, for example, [4].

If θ were known to be equal to θ^0, it is natural to consider a predictive distribution with density $g(z;\theta^0)$ and distribution function $G(z;\theta^0)$. The corresponding α-quantile $z_\alpha(\theta^0)$, specified as the solution to $G\{z_\alpha(\theta^0);\theta^0\} = \alpha$, satisfies (1) exactly. When θ is unknown, there are some special cases where there is an exact solution to (1), but this usually relies on the existence of a suitable pivotal function, that is a function of Y and Z whose distribution is free of θ. However, since this is the exception rather than the rule, we look for approximate solutions and, in particular, we aim at introducing a predictive distribution such that the corresponding α-quantiles fulfill (1) almost exactly, for each $\alpha \in (0, 1)$.

2.1 Plug-in predictive procedures and improvements

The plug-in, or estimative, approach to prediction consists of estimating θ^0 with a suitable estimator, usually the maximum likelihood estimator based on Y, namely $\widehat{\theta} = \widehat{\theta}(Y)$, and using $G(z;\widehat{\theta})$ and $g(z;\widehat{\theta})$ as predictive distribution and density functions, respectively. The corresponding α-quantile, $z_\alpha(\widehat{\theta})$, is called plug-in or estimative prediction limit. It is well-known that its coverage probability differs from α by a term usually of order $O(n^{-1})$ and prediction statements may be rather inaccurate for a small n. In fact, this naive solution underestimates the additional uncertainty introduced by assuming $\theta = \widehat{\theta}$.

By means of asymptotic calculations, [4] and [17] obtain an explicit expression for the $O(n^{-1})$ coverage error term. Since the coverage probability of a given prediction limit $c_\alpha(Y)$ may be rewritten as $P_{Y,Z}\{Z \le c_\alpha(Y);\theta^0\} = E_Y\{G(c_\alpha(Y);\theta^0)\}$, where the expectation is under the true distribution of Y, they prove that

$$P_{Y,Z}\{Z \le z_\alpha(\widehat{\theta});\theta^0\} \doteq \alpha + Q(z_\alpha(\theta^0);\theta^0). \tag{2}$$

Here

$$\begin{aligned} Q(z;\theta^0) = & -b_r(\theta^0)G_r(z;\theta^0) - (1/2)\, i^{rs}(\theta^0)\, G_{rs}(z;\theta^0) \\ & + i^{rs}(\theta^0)G_r(z;\theta^0)\ell_s(\theta^0;z), \end{aligned} \tag{3}$$

where $b_r(\theta^0)$ is the $O(n^{-1})$ bias term of r-th component of the maximum likelihood estimator $\widehat{\theta}$, while $i^{rs}(\theta^0)$, $r, s = 1, \ldots, d$, is the (r, s)-element of the inverse of the expected information matrix based on Y. $G_r(z;\theta^0)$ and $G_{rs}(z;\theta^0)$, $r, s = 1, \ldots, d$, are the first and the second partial derivatives of $G(z;\theta)$ with respect to the corresponding components of vector θ, evaluated at $\theta = \theta^0$, and $\ell_r(\theta^0; z)$, $r = 1, \ldots, d$, is $\partial\ell(\theta; z)/\partial\theta_r$, evaluated at $\theta = \theta^0$, where $\ell(\theta; z) = \log g(z;\theta)$. Hereafter, we use index notation and the Einstein summation convention, so that summation is implicitly understood if an index occurs more than once in a summand. Indeed, β_r, $r = 1, \ldots, d$, defines the r-th element of a d-dimensional vector β and the symbol $\doteq$ indicates that the equality holds up to terms of order $O(n^{-1})$ or $O_p(n^{-1})$, depending on the context.

It is quite easy to define a prediction limit, specified as a modification of the plug-in limit, so that the $O(n^{-1})$ error term in (2) disappears. This improved prediction limit is

$$z_\alpha^\dagger(Y) = z_\alpha(\widehat{\theta}) - \frac{Q(z_\alpha(\widehat{\theta});\widehat{\theta})}{g(z_\alpha(\widehat{\theta});\widehat{\theta})},$$

which is, up to terms of order $O(n^{-1})$, the α-quantile of a predictive distribution with

$$g^{\dagger}(z;Y) = g(z;\widehat{\theta})\left\{1 + \frac{\partial Q(z;\widehat{\theta})/\partial z}{g(z;\widehat{\theta})}\right\},$$
$$G^{\dagger}(z;Y) = G(z;\widehat{\theta}) + Q(z;\widehat{\theta}),$$

as the associated density and distribution functions, respectively. Note that both $G^{\dagger}(z;Y)$ and $g^{\dagger}(z;Y)$ are modifications of plug-in quantities as well.

Recently, [18] has defined a useful, simplified form for the predictive distribution function giving the improved prediction limit $z_{\alpha}^{\dagger}(Y)$. Alternative procedures for improving the plug-in prediction limit are proposed in [10] and [13]. The first involves a suitable simulation-based bootstrap calibration procedure, while the second is found on approximate pivotal quantities.

2.2 Profile predictive likelihood and modifications

A natural formulation of the likelihood principle for prediction state that all evidence about the unknown quantities z and θ is contained in the joint likelihood function $L(z,\theta;y) = p(y,z;\theta)$. Consequently, a predictive likelihood for z is a function obtained by eliminating the nuisance parameter θ from $L(z,\theta;y)$. Here, we do not present all the different notions of predictive likelihood but we concentrate on those obtained using maximisation and, in particular, we will consider the profile predictive likelihood and some modified versions.

The profile predictive likelihood (see [16]) is the predictive analogue of the (parametric) profile likelihood for an interest parameter, when nuisance parameters are present. It is defined as

$$L_p(z;y) = p(y,z;\widehat{\theta}_z) = g(z;\widehat{\theta}_z)p(y;\widehat{\theta}_z), \tag{4}$$

with $p(y;\widehat{\theta}_z) = \prod_{i=1}^{n} f(y_i;\widehat{\theta}_z)$. This differs from the plug-in approach outlined in Section 2.1 in that the unknown parameter is substituted by the constrained maximum likelihood estimator $\widehat{\theta}_z = \widehat{\theta}_z(Y,z)$, specified as the value of θ that maximises the joint likelihood $L(z,\theta;Y)$. Although simple and, in some sense appealing, this notion suffers from the same drawbacks as the plug-in solution, since it does not adequately account for the uncertainty in θ, while doing predictive inference on z.

Alternative predictive likelihoods, specified as modifications of $L_p(z;y)$, were suggested in [9] and [8]. In particular, Davison's proposal corresponds to an approximation, based on Laplace's integral formula, to the Bayesian predictive density when the contribution of the prior information is weak or flat priors are in fact considered. Thus, whenever terms related to the prior distribution are omitted in the approximation, we obtain Davison's predictive likelihood

$$L_d(z;y) = L_p(z;y)\frac{|\,J(\widehat{\theta})\,|^{1/2}}{p(y;\widehat{\theta})\,|\,J^z(\widehat{\theta}_z)\,|^{1/2}} \propto L_p(z;y)\,|\,J^z(\widehat{\theta}_z)\,|^{-1/2}, \tag{5}$$

where $|\,M\,|$ denotes the determinant of matrix M. Indeed, $J(\widehat{\theta})$ is the observed information matrix, computed from the log-likelihood $\ell(\theta;y) = \log p(y;\theta)$ and evaluated at $\theta = \widehat{\theta}$, with (r, s)-element $j_{rs}(\theta) = -\partial^2 \ell(\theta; y)/\partial\theta_r \partial\theta_s, r, s = 1, \ldots, d$, and $J^z(\widehat{\theta}_z)$ is the observed information matrix, computed from the joint log-likelihood $\ell(\theta; y, z) = \log p(y, z; \theta)$ and evaluated at $\theta = \widehat{\theta}_z$, with (r, s)-element $j^z_{rs}(\theta) = -\partial^2 \ell(\theta; y, z)/\partial\theta_r \partial\theta_s$, $r, s = 1, \ldots, d$.

Instead, the solution proposed in [8] is the predictive analogue of the modified profile likelihood introduced by [1] in parametric inference. It is called modified profile predictive likelihood and it is given by

$$L_m(z; y) = L_p(z; y) \mid J^z(\widehat{\theta}_z) \mid^{-1/2} \| \partial\widehat{\theta}/\partial\widehat{\theta}_z \|, \tag{6}$$

where $\| M \|$ denotes the absolute value of the determinant of matrix M and $\partial\widehat{\theta}/\partial\widehat{\theta}_z$ is the matrix of partial derivatives with (r, s)-element $\partial\widehat{\theta}_r/\partial\widehat{\theta}_{z,s}$, $r, s = 1, \ldots, d$. Function $L_m(z; y)$ is obtained as an approximation of the conditional density of Z given $\widehat{\theta}_z$, assuming that the transformation $(z, \widehat{\theta}) \to (z, \widehat{\theta}_z)$ is one-to-one.

It is known that both $L_p(z; y)$ and $L_m(z; y)$ are invariant under a one-to-one reparametrisation of the model, while $L_d(z; y)$ is not parameter invariant in general. Indeed, $L_m(z; y)$ cannot be considered when $\widehat{\theta}$ is not a function of $(z, \widehat{\theta}_z)$, as, for example, when $Y_1, \ldots, Y_n, Z$ follows a uniform distribution on $[0, \theta]$. Besides considering these relevant features, we aim to compare plug-in and likelihood based predictive solutions by analysing the coverage properties of the prediction intervals they generate.

3 Likelihood-based predictive distributions and prediction limits

One important use of predictive likelihoods is for constructing prediction limits. Furthermore, the evaluation of their coverage probabilities constitutes a way to compare different predictive procedures. In this section, the predictive likelihoods previously recalled are normalised to be probability distributions in z and the corresponding prediction limits are explicitly derived as quantiles. Moreover, for each limit, we compute the coverage probability.

We will show that, given a predictive likelihood $L(z; y)$, the associated predictive density and distribution functions are suitable modifications of those given by the plug-in, so that

$$g_L(z; Y) \doteq g(z; \widehat{\theta})\{1 + S_L(z; \widehat{\theta})\}$$
$$G_L(z; Y) \doteq G(z; \widehat{\theta}) + R_L(z; \widehat{\theta}),$$

where the $O(n^{-1})$ modifying terms are such that

$$S_L(z; \widehat{\theta}) = \frac{\partial R_L(z; \widehat{\theta})/\partial z}{g(z; \widehat{\theta})}, \quad R_L(z; \widehat{\theta}) = \int_{-\infty}^{z} S_L(t; \widehat{\theta}) g(t; \widehat{\theta}) dt.$$

To the relevant order of approximation, the corresponding α-prediction limit turns out to be

$$z^L_\alpha(Y) \doteq z_\alpha(\widehat{\theta}) - \frac{R_L(z_\alpha(\widehat{\theta}); \widehat{\theta})}{g(z_\alpha(\widehat{\theta}); \widehat{\theta})}. \tag{7}$$

Indeed, from (2), it is easy to state that the associated coverage probability is

$$P_{Y,Z}\{Z \leq z_\alpha^L(Y); \theta^0\} \doteq \alpha + Q(z_\alpha(\theta^0); \theta^0) - R_L(z_\alpha(\theta^0); \theta^0), \quad (8)$$

with $Q(z_\alpha(\theta^0); \theta^0)$ given by (3) evaluated at $z = z_\alpha(\theta^0)$.

3.1 Probability distributions from predictive likelihoods

Let us assume that $\widehat{\theta}$ and $\widehat{\theta}_z$ are solutions, with respect to θ, to the associated likelihood equations, namely, $\ell_r(\theta; Y) = 0$ and $\ell_r(\theta; Y) + \ell_r(\theta; z) = 0$, $r = 1, \ldots, d$, with $\ell_r(\theta; Y) = \partial\ell(\theta; Y)/\partial\theta_r$. Under regularity conditions, $\widehat{\theta}$ is a consistent estimator of θ^0, the maximiser of $E_Y\{\ell(\theta; Y)\}$ with respect to θ. It may be shown (see, for example, [9]) that for, $r = 1, \ldots, d$,

$$\widehat{\theta}_{z,r} \doteq \widehat{\theta}_r + j^{rs}(\widehat{\theta})\,\ell_s(\widehat{\theta}; z), \quad (9)$$

where $j^{rs}(\widehat{\theta})$, $r, s = 1, \ldots, d$, is the (r, s)-element of the inverse of the observed information matrix, based on Y and evaluated at $\theta = \widehat{\theta}$, which can be substituted by its expected counterpart $i^{rs}(\widehat{\theta})$ without changing the order of approximation.

Let us consider the profile predictive likelihood defined by (4). By means of a straightforward asymptotic expansion around $\widehat{\theta}_z = \widehat{\theta}$, we have that

$$\begin{aligned} L_p(z; y) &\doteq g(z; \widehat{\theta})p(y; \widehat{\theta}) + (\widehat{\theta}_z - \widehat{\theta})_r\{\partial L_p(z; y)/\partial\widehat{\theta}_{z,r}\}|_{\widehat{\theta}_z=\widehat{\theta}} \\ &\quad + (1/2)(\widehat{\theta}_z - \widehat{\theta})_{rs}\{\partial^2 L_p(z; y)/\partial\widehat{\theta}_{z,r}\,\partial\widehat{\theta}_{z,s}\}|_{\widehat{\theta}_z=\widehat{\theta}}, \end{aligned}$$

where $(\widehat{\theta}_z - \widehat{\theta})_{rs} = (\widehat{\theta}_z - \widehat{\theta})_r(\widehat{\theta}_z - \widehat{\theta})_s$. Retaining only the terms of order $O_p(n^{-1})$ and using relation (9), we obtain

$$L_p(z; y) \doteq g(z; \widehat{\theta})p(y; \widehat{\theta})\{1 + (1/2)i^{rs}(\widehat{\theta})\,\ell_r(\widehat{\theta}; z)\ell_s(\widehat{\theta}; z)\}, \quad (10)$$

with the expected information matrix considered instead of the observed information matrix. Thus, neglecting terms of order $o_p(n^{-1})$, $L_p(z; y)$ can be easily normalised with respect to z giving

$$g_p(z; Y) \doteq g(z; \widehat{\theta})[1 + (1/2)i^{rs}(\widehat{\theta})\{\ell_r(\widehat{\theta}; z)\ell_s(\widehat{\theta}; z) - c_{rs}(\widehat{\theta})\}],$$

with

$$c_{rs}(\widehat{\theta}) = \int_{-\infty}^{+\infty} \ell_r(\widehat{\theta}; z)\ell_s(\widehat{\theta}; z)g(z; \widehat{\theta})dz. \quad (11)$$

The associated distribution function is

$$G_p(z; Y) \doteq G(z; \widehat{\theta}) + (1/2)i^{rs}(\widehat{\theta})\{c_{rs}(\widehat{\theta}; z) - c_{rs}(\widehat{\theta})G(z; \widehat{\theta})\},$$

with

$$c_{rs}(\widehat{\theta}; z) = \int_{-\infty}^{z} \ell_r(\widehat{\theta}; t)\ell_s(\widehat{\theta}; t)g(t; \widehat{\theta})dt. \quad (12)$$

Note that $c_{rs}(\widehat{\theta}; +\infty) = c_{rs}(\widehat{\theta})$. Whenever the above integrals do not present an explicit solution, standard numerical approximation methods are considered.

Davison's predictive likelihood, as specified by (5), is a modification of $L_p(z; y)$. Thus, using relations (10) and (17), we obtain

$$L_d(z; y) \doteq g(z; \widehat{\theta})p(y; \widehat{\theta}) \mid J(\widehat{\theta}) \mid^{-1/2} [1 + (1/2)i^{rs}(\widehat{\theta})\{\ell_r(\widehat{\theta}; z)\ell_s(\widehat{\theta}; z) + \ell_{rs}(\widehat{\theta}; z)\} + (1/2)i^{rs}(\widehat{\theta})i^{tu}(\widehat{\theta})\ell_{rtu}(\widehat{\theta}; y)\ell_s(\widehat{\theta}; z)], \quad (13)$$

where $\ell_{rtu}(\widehat{\theta}; y)$ is $\partial^3\ell(\theta; y)/\partial\theta_r\partial\theta_t\partial\theta_u$, $r, t, u = 1, \ldots, d$, evaluated at $\theta = \widehat{\theta}$. Recalling the balance relations

$$\int_{-\infty}^{+\infty} \ell_r(\theta; z)g(z; \theta)dz = 0,$$

$$\int_{-\infty}^{+\infty} \ell_r(\theta; z)\ell_s(\theta; z)g(z; \theta)dz = -\int_{-\infty}^{+\infty} \ell_{rs}(\theta; z)g(z; \theta)dz,$$

a normalised version for $L_d(z; y)$ is readily available and it corresponds to

$$g_d(z; Y) \doteq g(z; \widehat{\theta})[1 + (1/2)i^{rs}(\widehat{\theta})\{\ell_r(\widehat{\theta}; z)\ell_s(\widehat{\theta}; z) + \ell_{rs}(\widehat{\theta}; z)\} + (1/2)i^{rs}(\widehat{\theta})i^{tu}(\widehat{\theta})\ell_{rtu}(\widehat{\theta}; y)\ell_s(\widehat{\theta}; z)].$$

The associated distribution function is

$$G_d(z; Y) \doteq G(z; \widehat{\theta}) + (1/2)i^{rs}(\widehat{\theta})G_{rs}(z; \widehat{\theta}) + (1/2)i^{rs}(\widehat{\theta})i^{tu}(\widehat{\theta})\ell_{rtu}(\widehat{\theta}; y)G_s(z; \widehat{\theta}),$$

where we implicitly assume that differentiation and integration may be interchanged so that

$$G_r(z; \theta) = \int_{-\infty}^{z} \ell_r(\theta; t)g(t; \theta)dt,$$

$$G_{rs}(z; \theta) = \int_{-\infty}^{z} \{\ell_r(\theta; t)\ell_s(\theta; t) + \ell_{rs}(\theta; t)\}g(z; \theta)dt.$$

Finally, let us consider the modified profile predictive likelihood defined by (6). Since it is a modification of $L_d(z; y)$, using relations (13) and (19), we state that

$$L_m(z; y) \doteq g(z; \widehat{\theta})p(y; \widehat{\theta}) \mid J(\widehat{\theta}) \mid^{-1/2} [1 + (1/2)i^{rs}(\widehat{\theta})\{\ell_r(\widehat{\theta}; z)\ell_s(\widehat{\theta}; z) - \ell_{rs}(\widehat{\theta}; z)\} - (1/2)i^{rs}(\widehat{\theta})i^{tu}(\widehat{\theta})\{\ell_{rtu}(\widehat{\theta}; y) + 2\ell_{rt,u}(\widehat{\theta}; y)\}\ell_s(\widehat{\theta}; z)],$$

with $\ell_{rt,u}(\widehat{\theta}; y)$ given by $\partial^3\ell(\theta; \widehat{\theta}, a)/\partial\theta_r\partial\theta_t\partial\widehat{\theta}_u$, $r, t, u = 1, \ldots, d$, evaluated at $\theta = \widehat{\theta}$. Here, we consider the log-likelihood function in the form $\ell(\theta; \widehat{\theta}, a) = \ell(\theta; y)$, with $(\widehat{\theta}, a)$ a sufficient reduction of data y, $a = a(y)$ being an exact or approximate ancillary statistic. A normalised version for $L_m(z; y)$ is

$$g_m(z; Y) \doteq g(z; \widehat{\theta})[1 + (1/2)i^{rs}(\widehat{\theta})\{\ell_r(\widehat{\theta}; z)\ell_s(\widehat{\theta}; z) - \ell_{rs}(\widehat{\theta}; z) - 2c_{rs}(\widehat{\theta})\} - (1/2)i^{rs}(\widehat{\theta})i^{tu}(\widehat{\theta})\{\ell_{rtu}(\widehat{\theta}; y) + 2\ell_{rt,u}(\widehat{\theta}; y)\}\ell_s(\widehat{\theta}; z)],$$

while the associated distribution function is

$$G_m(z; Y) \doteq G(z; \widehat{\theta}) - (1/2)i^{rs}(\widehat{\theta})G_{rs}(z; \widehat{\theta}) + i^{rs}(\widehat{\theta})\{c_{rs}(\widehat{\theta}; z) - c_{rs}(\widehat{\theta})G(z; \widehat{\theta})\}$$
$$-(1/2)i^{rs}(\widehat{\theta})i^{tu}(\widehat{\theta})\{\ell_{rtu}(\widehat{\theta}; y) + 2\ell_{rt,u}(\widehat{\theta}; y)\}G_s(z; \widehat{\theta}),$$

with $c_{rs}(\widehat{\theta})$ and $c_{rs}(\widehat{\theta}; z)$ given by (11) and (12).

Note that none of the distributions obtained from predictive likelihoods corresponds to the improved predictive distribution outlined in Section 2.1. In particular, all the additive modifying terms in the predictive distribution functions based on likelihood fail to consider both function $\ell_s(\widehat{\theta}; z)$ and the estimated first-order bias term of the maximum likelihood estimator. The r-th bias term may be approximated as $b_r(\widehat{\theta}) = -(1/2)i^{rs}(\widehat{\theta})i^{tu}(\widehat{\theta})\ell_{r,tu}(\widehat{\theta}; y)$, $r = 1, \ldots, d$, where $\ell_{r,tu}(\widehat{\theta}; y)$, $r, t, u = 1, \ldots, d$, is $\partial^3\ell(\theta; \widehat{\theta}, a)/\partial\theta_r\partial\widehat{\theta}_t\partial\widehat{\theta}_u$, evaluated at $\theta = \widehat{\theta}$ (see [3], Section 6.4).

3.2 Prediction limits and coverage probabilities

Using the results obtained in the previous section, it is almost immediate to specify the prediction limits generated by $L_p(z; y)$, $L_d(z; y)$ and $L_m(z; y)$. As a matter of fact, the corresponding α-prediction limits, namely $z_\alpha^p(Y)$, $z_\alpha^d(Y)$ and $z_\alpha^m(Y)$, are given by equation (7) with $L = p, d, m$, where

$$R_p(z_\alpha(\widehat{\theta}); \widehat{\theta}) = (1/2)i^{rs}(\widehat{\theta})\{c_{rs}(\widehat{\theta}; z) - c_{rs}(\widehat{\theta})G(z; \widehat{\theta})\},$$
$$R_d(z_\alpha(\widehat{\theta}); \widehat{\theta}) = (1/2)\{i^{rs}(\widehat{\theta})G_{rs}(z; \widehat{\theta}) + i^{rs}(\widehat{\theta})i^{tu}(\widehat{\theta})\ell_{rtu}(\widehat{\theta}; y)G_s(z; \widehat{\theta})\}$$

and

$$R_m(z_\alpha(\widehat{\theta}); \widehat{\theta}) = -(1/2)i^{rs}(\widehat{\theta})G_{rs}(z; \widehat{\theta}) + i^{rs}(\widehat{\theta})\{c_{rs}(\widehat{\theta}; z) - c_{rs}(\widehat{\theta})G(z; \widehat{\theta})\}$$
$$-(1/2)i^{rs}(\widehat{\theta})i^{tu}(\widehat{\theta})\{\ell_{rtu}(\widehat{\theta}; y) + 2\ell_{rt,u}(\widehat{\theta}; y)\}G_s(z; \widehat{\theta}).$$

Furthermore, by means of equation (8), it is possible to compute the coverage probabilities of $z_\alpha^p(Y)$, $z_\alpha^d(Y)$ and $z_\alpha^m(Y)$, up to terms of order $O(n^{-1})$. Since, $Q(z_\alpha(\theta^0); \theta^0)$ differs from $R_L(z_\alpha(\widehat{\theta}); \widehat{\theta})$, $L = p, d, m$, evaluated at $\widehat{\theta} = \theta^0$, we may conclude that prediction limits based on predictive likelihoods do not present high order coverage accuracy; their coverage probabilities differ from the target value α by terms usually of order $O(n^{-1})$.

In spite of these negative theoretical results, we find that at least in the same specific applications, such as those presented in the following section, these prediction limits improve on those based on the plug-in solution as well. Although the asymptotic order is not reduced, the first-order coverage error term turns out to be uniformly smaller, as confirmed by explicit calculations and evaluations using simulation-based procedures.

4 Examples

In this final section we present two simple applications related to the Gaussian model. Let us assume that $Y_1, \ldots, Y_n, Z_1, \ldots, Z_m$, $n \geq 1$, $m \geq 1$, are independently $N(\mu, \sigma^2)$ distributed with $\theta = (\mu, \sigma^2)$ unknown. Whenever $m = 1$, it is well-known that the future random variable $Z = Z_1$ is such that $(Z - \widehat{\mu})/\sqrt{\widehat{\sigma}^2(n+1)/(n-1)}$

follows a t-distribution with $n-1$ degrees of freedom, with $\widehat{\mu}$ and $\widehat{\sigma}^2$ the maximum likelihood estimators for μ and σ^2 based on Y; here, $\widehat{\theta} = (\widehat{\mu}, \widehat{\sigma}^2)$. Thus, in this simple case, there is an exact solution to the prediction problem based on a pivotal quantity. In the following two examples Z is defined, respectively, as the sum and the maximum of the m future independent observations $Z_1, \ldots, Z_m$. In these cases it does not seem possible to define a pivotal quantity for specifying exact prediction limits and we focus on the approximate solutions outlined in the paper.

Some useful results are now reviewed. Let $\phi = \phi(z; \mu, \sigma^2)$ and $\Phi = \Phi(z; \mu, \sigma^2)$ be the density and the distribution function of an $N(\mu, \sigma^2)$ distribution. Using the properties of Hermite polynomials (see [2], Section 1.6), we have

$$\Phi_\mu = \Phi_\mu(z; \mu, \sigma^2) = -\phi,$$
$$\Phi_{\sigma^2} = \Phi_{\sigma^2}(z; \mu, \sigma^2) = -\frac{z-\mu}{2\sigma^2}\,\phi,$$
$$\Phi_{\mu\mu} = \Phi_{\mu\mu}(z; \mu, \sigma^2) = -\frac{z-\mu}{\sigma^2}\,\phi,$$
$$\Phi_{\sigma^2\sigma^2} = \Phi_{\sigma^2\sigma^2}(z; \mu, \sigma^2) = -\left\{\frac{(z-\mu)^3}{4\sigma^6} - 3\frac{(z-\mu)}{4\sigma^4}\right\}\phi,$$

where the subscripts μ and σ^2 indicate differentiation with respect to the corresponding parameters. Indeed, the first-order bias terms of the maximum likelihood estimators are $b_\mu = 0$ and $b_{\sigma^2} = -\sigma^2/n$ and the non-null terms of the inverse of the expected information matrix are $i^{\mu\mu} = \sigma^2/n$ and $i^{\sigma^2\sigma^2} = 2\sigma^4/n$. Moreover, $i^{r\sigma^2}(\widehat{\theta})i^{tu}(\widehat{\theta})\ell_{rtu}(\widehat{\theta}; y) = 10\widehat{\sigma}^2/n$ and $i^{r\sigma^2}(\widehat{\theta})i^{tu}(\widehat{\theta})\ell_{rt,u}(\widehat{\theta}; y) = -6\widehat{\sigma}^2/n$.

4.1 Prediction limits for the sum of future Gaussian observations

Let us consider $Z = \sum_{j=1}^m Z_j$, following a $N(m\mu, m\sigma^2)$ distribution. Since $G(z;\theta) = \Phi(z; m\mu, m\sigma^2)$ and $g(z;\theta) = \phi(z; m\mu, m\sigma^2)$, we have that $G_r(z;\theta) = m\Phi_r$ and $G_{rr}(z;\theta) = m^2\Phi_{rr}$, with $r = \mu, \sigma^2$ and Φ_r, Φ_{rr} defined previously. Indeed, the constrained maximum likelihood estimator for θ is $\widehat{\theta}_z = (\widehat{\mu}_z, \widehat{\sigma}_z^2)$ with

$$\widehat{\mu}_z \doteq \widehat{\mu} + (z - m\widehat{\mu})/n, \quad \widehat{\sigma}_z^2 \doteq \widehat{\sigma}^2 + \{(z - m\widehat{\mu})^2 - \widehat{\sigma}^2\}/n$$

and

$$\ell_\mu(\theta; z) = (z - m\mu)/\sigma^2, \qquad \ell_{\sigma^2}(\theta; z) = -(1/2)\sigma^{-2} + (z - m\mu)^2/(2m\sigma^4),$$
$$\ell_{\mu\mu}(\theta; z) = -m/\sigma^2, \qquad \ell_{\sigma^2\sigma^2}(\theta; z) = (1/2\sigma^4) - (z - m\mu)^2/(m\sigma^6).$$

The plug-in predictive distribution function is easily defined as $\Phi(z; m\widehat{\mu}, m\widehat{\sigma}^2)$ and the corresponding α-prediction limit is $z_\alpha(\widehat{\theta}) = m\widehat{\mu} + \sqrt{m}\widehat{\sigma}u_\alpha$, with u_α the α-quantile of an $N(0,1)$ distribution. Using algebra, it is quite easy to compute the improved predictive distribution function and those based on predictive likelihoods. In particular, since in this case we prove that $|\,\partial\widehat{\theta}/\partial\widehat{\theta}_z\,| \doteq 1 + (m+1)/n$, the normalisation of $L_d(z; y)$ and $L_m(z; y)$ gives, to the relevant order of approximation, the same predictive distribution. Then, we obtain

$$z_\alpha^\dagger(Y) \doteq z_\alpha(\widehat{\theta}) + \widehat{\sigma}\sqrt{m}\{u_\alpha^3 + (2m+3)u_\alpha\}/(4n)$$
$$z_\alpha^p(Y) \doteq z_\alpha(\widehat{\theta}) + \widehat{\sigma}\sqrt{m}\{u_\alpha^3 + (2m+1)u_\alpha\}/(4n)$$
$$z_\alpha^d(Y) \doteq z_\alpha^m(Y) \doteq z_\alpha(\widehat{\theta}) + \widehat{\sigma}\sqrt{m}\{u_\alpha^3 + (2m+7)u_\alpha\}/(4n).$$

The associate coverage probabilities do not depend on θ^0 and correspond to

$$P_{Y,Z}\{Z \leq z_\alpha(\widehat{\theta}); \theta^0\} \doteq \alpha - \{u_\alpha^3 + (2m+3)u_\alpha\}\phi(u_\alpha)/(4n)$$
$$P_{Y,Z}\{Z \leq z_\alpha^p(Y); \theta^0\} \doteq \alpha - u_\alpha\phi(u_\alpha)/(2n),\ P_{Y,Z}\{Z \leq z_\alpha^d(Y); \theta^0\} \doteq \alpha + u_\alpha\phi(u_\alpha)/n,$$

where $\phi(u_\alpha) = \phi(u_\alpha; 0, 1)$, while for the improved prediction limit the $O(n^{-1})$ error term vanishes. Note that the absolute value of the coverage error of the plug-in prediction limit increases with m, so that its accuracy could be very misleading, as illustrated by Figure 1. Indeed, the accuracy of $z_\alpha^p(Y)$ is better than that of $z_\alpha^d(Y)$, for each $\alpha \in (0, 1)$. The results of a simulation study, not presented in this paper, emphasise that the improved prediction limit performs uniformly better than alternative proposals. Indeed, the plug-in solution is the worst, while the profile prediction limit is superior to Davison's.

By means of a suitable Cornish-Fisher type expansion we obtain $u_\alpha = t_\alpha - \frac{1}{4}(t_\alpha^3 + t_\alpha)n^{-1} + o(n^{-1})$, where t_α is the α-quantile of a t distribution with $n - q$ degrees of freedom, with $q < n$ fixed. Using this expansion, we prove that, for $m = 1$, the improved prediction limit coincides, up to terms of order $O_p(n^{-1})$, to the exact solution recalled at the beginning of Section 4. Moreover, in accordance with [5], we find that, to the relevant order of approximation, $z_\alpha^p(Y)$ equals the α-quantile of a random variable X such that $(X - m\widehat{\mu})/\sqrt{m\widehat{\sigma}^2\{1 + (m/n)\}}$ follows a t distribution with n degrees of freedom, while $z_\alpha^d(Y)$ corresponds to the α-quantile of a random variable X such that $(X - m\widehat{\mu})/\sqrt{m\widehat{\sigma}^2\{1 + (m+3)/n\}}$ follows a t distribution with $n - 3$ degrees of freedom.

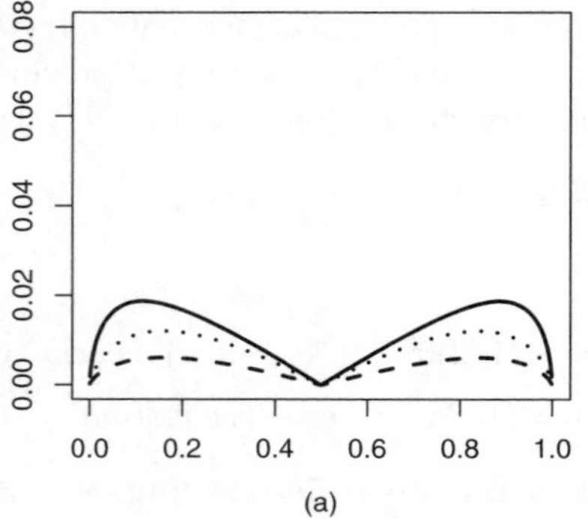

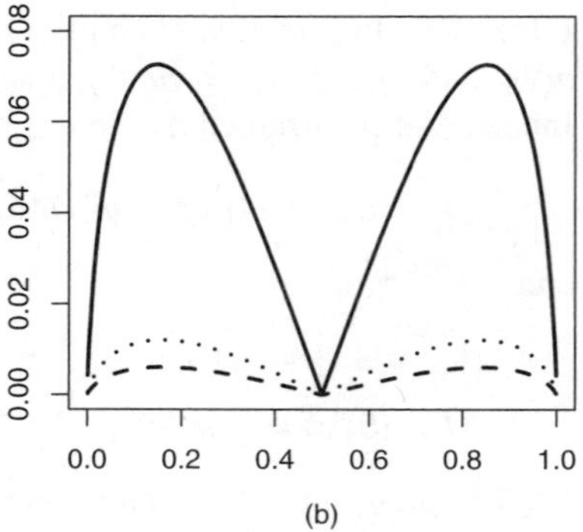

Fig. 1. Absolute value of the $O(n^{-1})$ coverage error term of the plug-in (solid line), profile (dashed line) and Davison's (dotted line) prediction limits, based on $n = 20$ independent observations from a Gaussian distribution. Prediction for the sum of m independent observations from the same Gaussian distribution, with **(a)** $m = 1$ and **(b)** $m = 10$

4.2 Prediction limits for the maximum of future Gaussian observations

Let us consider $Z = \max\{Z_1, \ldots, Z_m\}$. Since $G(z;\theta) = \Phi(z;\mu,\sigma^2)^m$ and $g(z;\theta) = m\phi(z;\mu,\sigma^2)\Phi(z;\mu,\sigma^2)^{m-1}$, we have $G_r(z;\theta) = m\Phi^{m-1}\Phi_r$ and $G_{rr}(z;\theta) = m(m-1)\Phi^{m-2}\Phi_r^2 + m\Phi^{m-1}\Phi_{rr}$, with $r = \mu, \sigma^2$. Indeed, the constrained maximum likelihood estimators are

$$
\begin{aligned}
\widehat{\mu}_z &\doteq \widehat{\mu} + (z - \widehat{\mu})/n - (m-1)\widehat{\sigma}^2\widehat{\phi}/(n\widehat{\Phi}),\\
\widehat{\sigma}_z^2 &\doteq \widehat{\sigma}^2 + \{(z-\widehat{\mu})^2 - \widehat{\sigma}^2\}/n - (m-1)\widehat{\sigma}^2(z-\widehat{\mu})\widehat{\phi}/(n\widehat{\Phi}),
\end{aligned}
$$

with $\widehat{\phi} = \phi(z;\widehat{\mu},\widehat{\sigma}^2)$, $\widehat{\Phi} = \Phi(z;\widehat{\mu},\widehat{\sigma}^2)$, and

$$
\begin{aligned}
\ell_\mu(\theta;z) &= (z-\mu)/\sigma^2 + (m-1)\Phi_\mu/\Phi,\\
\ell_{\sigma^2}(\theta;z) &= -(1/2)\sigma^{-2} + (z-\mu)^2/(2m\sigma^4) + (m-1)\Phi_{\sigma^2}/\Phi,\\
\ell_{\mu\mu}(\theta;z) &= -1/\sigma^2 + (m-1)\{\Phi_{\mu\mu}/\Phi - (\Phi_\mu/\Phi)^2\},\\
\ell_{\sigma^2\sigma^2}(\theta;z) &= (1/2\sigma^4) - (z-\mu)^2/(m\sigma^6) + (m-1)\{\Phi_{\sigma^2\sigma^2}/\Phi - (\Phi_{\sigma^2}/\Phi)^2\}.
\end{aligned}
$$

The plug-in predictive distribution function is easily defined as $\Phi(z;\widehat{\mu},\widehat{\sigma}^2)^m$ and the corresponding α-prediction limit can be alternatively specified as $z_\alpha(\widehat{\theta}) = \widehat{\mu} + \widehat{\sigma}u_{\alpha,m}$, with $u_{\alpha,m}$ the $\alpha^{1/m}$-quantile of an $N(0,1)$ distribution. With some algebra, we compute the improved predictive distribution function and those based on predictive likelihoods. Then, we obtain

$$
\begin{aligned}
z_\alpha^\dagger(Y) &\doteq z_\alpha(\widehat{\theta}) + \widehat{\sigma}\{u_{\alpha,m}^3 + 5u_{\alpha,m} - \widehat{\sigma}(m-1)(u_{\alpha,m}^2 + 2)\widehat{\phi}(z_\alpha(\widehat{\theta}))\alpha^{-1/m}\}/(4n)\\
z_\alpha^p(Y) &\doteq z_\alpha(\widehat{\theta}) - (1/2)h(\widehat{\theta})/\{m\widehat{\phi}(z_\alpha(\widehat{\theta}))\alpha^{(m-1)/m}\},\\
z_\alpha^d(Y) &\doteq z_\alpha(\widehat{\theta}) + \widehat{\sigma}\{u_{\alpha,m}^3 + 9u_{\alpha,m} - \widehat{\sigma}(m-1)(u_{\alpha,m}^2 + 2)\widehat{\phi}(z_\alpha(\widehat{\theta}))\alpha^{-1/m}\}/(4n)\\
z_\alpha^m(Y) &\doteq z_\alpha(\widehat{\theta}) - \widehat{\sigma}\{u_{\alpha,m}^3 - 3u_{\alpha,m} - \widehat{\sigma}(m-1)(u_{\alpha,m}^2 + 2)\widehat{\phi}(z_\alpha(\widehat{\theta}))\alpha^{-1/m}\}/(4n)\\
&\quad - h(\widehat{\theta})/\{m\widehat{\phi}(z_\alpha(\widehat{\theta}))\alpha^{(m-1)/m}\},
\end{aligned}
$$

where $\widehat{\phi}(z_\alpha(\widehat{\theta})) = \phi(z_\alpha(\widehat{\theta});\widehat{\mu},\widehat{\sigma}^2)$ and $h(\widehat{\theta}) = i^{rr}(\widehat{\theta})\{c_{rr}(\widehat{\theta};z_\alpha(\widehat{\theta})) - \alpha c_{rr}(\widehat{\theta})\}$, defined using (11) and (12), are computed numerically. The coverage probabilities are

$$
\begin{aligned}
P_{Y,Z}\{Z \le z_\alpha(\widehat{\theta});\theta^0\} &\doteq \alpha - m[\{u_{\alpha,m}^3 + 5u_{\alpha,m}\}\phi(u_{\alpha,m})\alpha^{(m-1)/m}\\
&\quad -(m-1)\{u_{\alpha,m}^2 + 2\}\phi(u_{\alpha,m})^2\alpha^{(m-2)/m}]/(4n)\\
P_{Y,Z}\{Z \le z_\alpha^p(Y);\theta^0\} &\doteq \alpha - m[\{u_{\alpha,m}^3 + 5u_{\alpha,m}\}\phi(u_{\alpha,m})\alpha^{(m-1)/m}\\
&\quad -(m-1)\{u_{\alpha,m}^2 + 2\}\phi(u_{\alpha,m})^2\alpha^{(m-2)/m}]/(4n) - (1/2)h(\theta^0)\\
P_{Y,Z}\{Z \le z_\alpha^d(Y);\theta^0\} &\doteq \alpha + m u_{\alpha,m}\phi(u_{\alpha,m})\alpha^{(m-1)/m}/n\\
P_{Y,Z}\{Z \le z_\alpha^m(Y);\theta^0\} &\doteq \alpha - m[\{u_{\alpha,m}^3 + u_{\alpha,m}\}\phi(u_{\alpha,m})\alpha^{(m-1)/m}\\
&\quad -(m-1)\{u_{\alpha,m}^2 + 2\}\phi(u_{\alpha,m})^2\alpha^{(m-2)/m}]/(2n) - h(\theta^0),
\end{aligned}
$$

where $\phi(u_{\alpha,m}) = \phi(u_{\alpha,m};0,1)$, while $h(\theta^0)$ is $h(\widehat{\theta})$ evaluated at $\widehat{\theta} = \theta^0$. We prove that $h(\theta^0)$ does not depend on θ^0, so that all these coverage probabilities do not vary with θ^0. For the improved prediction limit the $O(n^{-1})$ error term vanishes.

A simple simulation experiment, which is a part of a wider study not presented in the paper, confirms the superiority of improved and likelihood-based prediction methods over the plug-in procedure. Independent observations of size $n = 10, 25, 50$ are generated from a normal model with $\mu = 0$ and $\sigma^2 = 1$; the dimension m of the future sample ranges from 1 to 15. The case $m = 1$ corresponds to a single future normal observation. Table 1 gives estimates of the actual confidence level for alternative α-prediction limits, with $\alpha = 0.90, 0.95$; the corresponding estimated standard errors are lower than 0.004. These simulations, based on 10,000 replications, show that $z_\alpha^\dagger(Y)$ performs uniformly better than the plug-in prediction limit and it seems to give more stable results than those obtained using predictive likelihood methods.

Table 1. Coverage probabilities of α-prediction limits, with $\alpha = 0.90, 0.95$, based on plug-in, improved and likelihood-based procedures. Estimates based on 10,000 simulated samples of size $n = 10, 25, 50$ from an $N(0, 1)$ distribution. Prediction for the maximum of a future sample of size $m = 1, 5, 15$. Estimated standard errors lower than 0.004

n	m	$z_\alpha(\widehat{\theta})$	$z_\alpha^\dagger(Y)$	$z_\alpha^p(Y)$	$z_\alpha^d(Y)$	$z_\alpha^m(Y)$	$z_\alpha(\widehat{\theta})$	$z_\alpha^\dagger(Y)$	$z_\alpha^p(Y)$	$z_\alpha^d(Y)$	$z_\alpha^m(Y)$
		$\alpha = 0.90$					$\alpha = 0.95$				
10	1	0.867	0.898	0.889	0.912	0.912	0.915	0.947	0.939	0.957	0.957
	5	0.795	0.890	0.866	0.917	0.909	0.862	0.938	0.925	0.954	0.950
	15	0.739	0.884	0.845	0.917	0.893	0.812	0.933	0.913	0.953	0.941
25	1	0.888	0.901	0.897	0.909	0.909	0.938	0.949	0.946	0.955	0.955
	5	0.854	0.895	0.883	0.911	0.906	0.916	0.947	0.940	0.956	0.954
	15	0.832	0.895	0.875	0.912	0.900	0.894	0.945	0.932	0.958	0.948
50	1	0.889	0.897	0.895	0.902	0.902	0.942	0.947	0.946	0.952	0.952
	5	0.884	0.903	0.897	0.913	0.909	0.937	0.955	0.950	0.958	0.957
	15	0.870	0.902	0.891	0.913	0.905	0.927	0.949	0.944	0.956	0.952

Appendix

We derive the asymptotic expansions for $| J^z(\widehat{\theta}_z) |^{-1/2}$ and $| \partial\widehat{\theta}/\partial\widehat{\theta}_z |$, which are used for obtaining the results outlined in Section 3.1.

Since $J^z(\widehat{\theta}_z) = J(\widehat{\theta}_z) + K(\widehat{\theta}_z)$, where $K(\widehat{\theta}_z)$ is a matrix, with (r, s)-element $k_{rs}(\theta) = -\partial^2\ell(\theta; z)/\partial\theta_r\partial\theta_s$, $r, s = 1, \ldots, d$, computed at $\theta = \widehat{\theta}_z$, using standard matrix calculation rules, we have that

$$| J^z(\widehat{\theta}_z) |^{-1/2} = | J(\widehat{\theta}_z) |^{-1/2} | I_d + J(\widehat{\theta}_z)^{-1} K(\widehat{\theta}_z) |^{-1/2}, \tag{14}$$

where I_d is the identity matrix. Let us consider the following stochastic Taylor expansions around $\widehat{\theta}_z = \widehat{\theta}$

$$\mid J(\widehat{\theta}_z) \mid^{-1/2} \doteq \mid J(\widehat{\theta}) \mid^{-1/2} + (\widehat{\theta}_z - \widehat{\theta})_r \{\partial \mid J(\widehat{\theta}_z) \mid^{-1/2} / \partial \widehat{\theta}_{z,r}\}|_{\widehat{\theta}_z = \widehat{\theta}}$$
$$\doteq \mid J(\widehat{\theta}) \mid^{-1/2} [1 - (1/2)(\widehat{\theta}_z - \widehat{\theta})_r \{\partial \log \mid J(\widehat{\theta}_z) \mid / \partial \widehat{\theta}_{z,r}\}|_{\widehat{\theta}_z = \widehat{\theta}}].$$

Using relation (9), and recalling that $\partial \log \mid J(\theta) \mid / \partial \theta_r\} = -j^{tu}(\theta)\ell_{rtu}(\theta; y)$ (see, for example, [3], Section 5.7), gives

$$\mid J(\widehat{\theta}_z) \mid^{-1/2} \doteq \mid J(\widehat{\theta}) \mid^{-1/2} \{1 + (1/2) j^{rs}(\widehat{\theta}) j^{tu}(\widehat{\theta}) \ell_{rtu}(\widehat{\theta}; y) \ell_s(\widehat{\theta}; z)\}. \tag{15}$$

Moreover, since the (r, t)-element of $I_d + J(\widehat{\theta}_z)^{-1} K(\widehat{\theta}_z)$ is $\delta_{rt} + \gamma_{rt}$, $r, t = 1, \ldots, d$, with $\delta_{rr} = 1$, $\delta_{rt} = 0$, when $r \neq t$, and $\gamma_{rt} = O_p(n^{-1})$, we state that

$$\begin{aligned} \mid I_d + J(\widehat{\theta}_z)^{-1} K(\widehat{\theta}_z) \mid^{-1/2} &\doteq \{1 + tr(J(\widehat{\theta}_z)^{-1} K(\widehat{\theta}_z))\}^{-1/2} \\ &\doteq \{1 + \gamma_{rr}\}^{-1/2} \\ &= \{1 - j^{rs}(\widehat{\theta}_z) \ell_{rs}(\widehat{\theta}_z; z)\}^{-1/2} \\ &\doteq 1 + (1/2) j^{rs}(\widehat{\theta}) \ell_{rs}(\widehat{\theta}; z), \end{aligned} \tag{16}$$

where $tr(M)$ is the trace of a matrix M and $\widehat{\theta}$ is considered instead of $\widehat{\theta}_z$, without changing the order of approximation. Finally, substitution of (15) and (16) into (14), gives

$$\begin{aligned} \mid J^z(\widehat{\theta}_z) \mid^{-1/2} \doteq \mid J(\widehat{\theta}) \mid^{-1/2} \{&1 + (1/2) j^{rs}(\widehat{\theta}) j^{tu}(\widehat{\theta}) \ell_{rtu}(\widehat{\theta}; y) \ell_s(\widehat{\theta}; z) \\ &+ (1/2) j^{rs}(\widehat{\theta}) \ell_{rs}(\widehat{\theta}; z)\}. \end{aligned} \tag{17}$$

In order to obtain an asymptotic expansion for $\mid \partial\widehat{\theta} / \partial\widehat{\theta}_z \mid$, we consider the following relation:

$$\widehat{\theta}_r \doteq \widehat{\theta}_{z,r} - j^{rt}(\widehat{\theta}_z)\, \ell_t(\widehat{\theta}_z; z), \tag{18}$$

obtained from (9) by means of a suitable inversion procedure. Differentiation, with respect to $\widehat{\theta}_{z,s}$, of $\widehat{\theta}_r$ as specified by (18), gives

$$\begin{aligned} \partial\widehat{\theta}_r / \partial\widehat{\theta}_{z,s} \doteq\; &\delta_{rs} - j^{ru}(\widehat{\theta}) j^{tk}(\widehat{\theta}) \{\ell_{uks}(\widehat{\theta}; y) + \ell_{uk,s}(\widehat{\theta}; y)\} \ell_t(\widehat{\theta}; z) \\ &- j^{rt}(\widehat{\theta}) \ell_{ts}(\widehat{\theta}; z), \end{aligned}$$

where $\widehat{\theta}$ is substituted for $\widehat{\theta}_z$, without changing the order of approximation. In the calculations we consider that

$$\partial j^{rt}(\widehat{\theta}) / \partial\widehat{\theta}_s = -j^{ru}(\widehat{\theta}) j^{tk}(\widehat{\theta}) \{\partial j_{uk}(\widehat{\theta}) / \partial\widehat{\theta}_s\}$$

and we take into account that $\ell(\widehat{\theta}; y) = \ell(\widehat{\theta}; \widehat{\theta}, a)$. Thus, the (r, s)-element of matrix $\partial\widehat{\theta} / \partial\widehat{\theta}_z$ is expressed as $\delta_{rs} + \gamma_{rs} + o_p(n^{-1})$, $r, s = 1, \ldots, d$, with $\gamma_{rs} = O_p(n^{-1})$, and we conclude that $\mid \partial\widehat{\theta} / \partial\widehat{\theta}_z \mid \doteq 1 + \gamma_{rr}$, namely

$$\begin{aligned} \mid \partial\widehat{\theta} / \partial\widehat{\theta}_z \mid \doteq\; &1 - j^{ru}(\widehat{\theta}) j^{tk}(\widehat{\theta}) \{\ell_{ukr}(\widehat{\theta}; y) + \ell_{uk,r}(\widehat{\theta}; y)\} \ell_r(\widehat{\theta}; z) \\ &- j^{rt}(\widehat{\theta}) \ell_{tr}(\widehat{\theta}; z). \end{aligned} \tag{19}$$

Whenever (19) is negative, we in fact consider its absolute value.

References

1. Barndorff-Nielsen, O.E.: On a formula for the distribution of the maximum likelihood estimator. Biometrika **70**, 343–365 (1983)
2. Barndorff-Nielsen, O.E., Cox, D.R.: Asymptotic Techniques for Use in Statistics. Chapman and Hall, London (1989)
3. Barndorff-Nielsen, O.E., Cox, D.R.: Inference and Asymptotics. Chapman and Hall, London (1994)
4. Barndorff-Nielsen, O.E., Cox, D.R.: Prediction and asymptotics. Bernoulli **2**, 319–340 (1996)
5. Bjørnstad, J.F.: Predictive likelihood: A review. Statist. Sci. **5**, 242–265 (1990)
6. Bjørnstad, J.F.: On the generalization of the likelihood function and the likelihood principle. J. Amer. Statist. Assoc. **91**, 791–806 (1996)
7. Butler, R.W.: Predictive likelihood inference with applications (with discussion). J. Roy. Statist. Soc. Ser. B **48**, 1–38 (1986)
8. Butler, R.W.: Approximate predictive pivots and densities. Biometrika **76**, 489–501 (1989)
9. Davison, A.C.: Approximate predictive likelihood. Biometrika **73**, 323–332 (1986)
10. Hall, P., Peng, L., Tajvidi, N.: On prediction intervals based on predictive likelihood or bootstrap methods. Biometrika **86**, 871–880 (1999)
11. Hinkley, D.: Predictive likelihood. Ann. Statist. **7**, 718–728 (1979)
12. Lauritzen, S.L.: Sufficiency, prediction and extreme models. Scand. J. Statist. **1**, 128–134 (1974)
13. Lawless, J.F., Fredette, M.: Frequentist prediction intervals and predictive distributions. Biometrika **92**, 529–542 (2005)
14. Lejeune, M., Faulkenberry, G.D.: A simple predictive density function. J. Amer. Statist. Assoc. **77**, 654–657 (1982)
15. Levy, M.S., Perng, S.K.: A maximum likelihood prediction function for the linear model with consistency results. Comm. Statist. A–Theory Methods **13**, 1257–1273 (1984)
16. Mathiasen, P.E.: Prediction functions. Scand. J. Statist. **6**, 1–21 (1979)
17. Vidoni, P.: A note on modified estimative prediction limits and distributions. Biometrika **85**, 949–953 (1998)
18. Vidoni, P.: Improved prediction intervals and distribution functions. Scand. J. Statist. **36**, 735–748 (2009)

Computer-intensive conditional inference

G. Alastair Young and Thomas J. DiCiccio

Abstract. Conditional inference is a fundamental part of statistical theory. However, exact conditional inference is often awkward, leading to the desire for methods which offer accurate approximations. Such a methodology is provided by small-sample likelihood asymptotics. We argue in this paper that simple, simulation-based methods also offer accurate approximations to exact conditional inference in multiparameter exponential family and ancillary statistic settings. Bootstrap simulation of the marginal distribution of an appropriate statistic provides a conceptually simple and highly effective alternative to analytic procedures of approximate conditional inference.

Key words: analytic approximation, ancillary statistic, Bartlett correction, bootstrap, conditional inference, exponential family, likelihood ratio statistic, stability

1 Introduction

Conditional inference has been, since the seminal work of Fisher [16], a fundamental part of the theory of parametric inference, but is a less established part of statistical practice.

Conditioning has two principal operational objectives: (i) the elimination of nuisance parameters; (ii) ensuring relevance of inference to an observed data sample, through the conditionality principle, of conditioning on the observed value of an ancillary statistic, when such a statistic exists. The concept of an ancillary statistic here is usually taken simply to mean one which is distribution constant. The former notion is usually associated with conditioning on sufficient statistics, and is most transparently and uncontroversially applied for inference in multiparameter exponential family models. Basu [7] provides a general and critical discussion of conditioning to eliminate nuisance parameters. The notion of conditioning to ensure relevance, together with the associated problem, which exercised Fisher himself (Fisher [17]), of recovering information lost when reducing the dimension of a statistical problem (say, to that of the maximum likelihood estimator, when this is not sufficient), is most transparent in transformation models, such as the location-scale model considered by Fisher [16].

Mantovan, P., Secchi, P. (Eds.): Complex Data Modeling and Computationally Intensive Statistical Methods

In some circumstances issues to do with conditioning are clear cut. Though most often applied as a slick way to establish independence between two statistics, Basu's Theorem (Basu [4]) shows that a boundedly complete sufficient statistic is independent of every distribution constant statistic. This establishes the irrelevance for inference of any ancillary statistic when a boundedly complete sufficient statistic exists.

In many other circumstances however, we have come to understand that there are formal difficulties with conditional inference. We list just a few. (1) It is well understood that conflict can emerge between conditioning and conventional measures of repeated sampling optimality, such as power. The most celebrated illustration is due to Cox [11]. (2) Typically there is arbitrariness on what to condition on. In particular, ancillary statistics are often not unique and a maximal ancillary may not exist. See, for instance, Basu [5, 6] and McCullagh [23]. (3) We must also confront the awkward mathematical contradiction of Birnbaum [9], which says that the conditionality principle, taken together with the quite uncontroversial sufficiency principle, imply acceptance of the likelihood principle of statistical inference, which is incompatible with the common methods of inference, such as calculation of p-values or construction of confidence sets, where we are drawn to the notion of conditioning.

Calculating a conditional sampling distribution is also typically not easy, and such practical difficulties, taken together with the formal difficulties with conditional inference, have led to much of modern statistical theory being based on notions of inference which automatically accommodate conditioning, at least to some high order of approximation. Of particular focus are methods which respect the conditionality principle without requiring explicit specification of the conditioning ancillary, and which therefore circumvent the difficulties associated with non-uniqueness of ancillaries.

Much attention in parametric theory now lies, therefore, in inference procedures which are stable, that is, which are based on a statistic which has, to some high order in the available data sample size, the same repeated sampling behaviour both marginally and conditional on the value of the appropriate conditioning statistic. The notion is that accurate approximation to an exact conditional inference can then be achieved by considering the marginal distribution of the stable statistic, ignoring the relevant conditioning. This idea is elegantly expressed for the ancillary statistic context by, for example, Barndorff-Nielsen and Cox [2, Section 7.2], Pace and Salvan [24, Section 2.8] and Severini [26, Section 6.4]. See also Efron and Hinkley [15] and Cox [12].

A principal approach to approximation of an intractable exact conditional inference by this route lies in developments in higher-order small-sample likelihood asymptotics, based on saddle point and related analytic methods. Book length treatments of this analytic approach are given by Barndorff-Nielsen and Cox [2] and Severini [26]. Brazzale *et al.* [10] demonstrate very convincingly how to apply these developments in practice. Methods have been constructed which automatically achieve, to a high order of approximation, the elimination of nuisance parameters which is desired in the exponential family setting, though focus has been predominantly on ancillary statistic models. Here, a key development concerns construction of adjusted forms of the signed root likelihood ratio statistic, which require specification of the ancillary statistic, but are distributed, conditionally on the ancillary, as $N(0, 1)$ to third order, $O(n^{-3/2})$, in the data sample size n. Normal approximation to the sampling distribution of the adjusted statistic therefore provides third-order approximation to exact conditional inference: see Barndorff-Nielsen [1]. Approximations which yield second-order conditional accuracy, that is, which approximate the exact conditional inference to an error of order $O(n^{-1})$, but which avoid specification of the ancillary statistic, are possible: Severini [26, Section 7.5] reviews such methods.

In the computer age, an attractive alternative approach to approximation of conditional inference uses marginal simulation, or 'parametric bootstrapping', of an appropriately chosen statistic to mimic its conditional distribution. The idea may be applied to approximate the conditioning that is appropriate to eliminate nuisance parameters in the exponential family setting, and can be used in ancillary statistic models, where specification of the conditioning ancillary statistic is certainly avoided.

Our primary purpose in this article is to review the properties of parametric bootstrap procedures in approximation of conditional inference. The discussion is phrased in terms of the inference problem described in Section 2. Exponential family and ancillary statistics models are described in Section 3. Key developments in analytic approximation methods are described in Section 4. Theoretical properties of the parametric bootstrap approach are described in Section 5, where comparisons are drawn with analytic approximation methods. A set of numerical examples are given in Section 6, with concluding remarks in Section 7.

2 An inference problem

We consider the following inference problem. Let $Y = \{Y_1, \ldots, Y_n\}$ be a random sample from an underlying distribution $F(y; \eta)$, indexed by a d-dimensional parameter η, where each Y_i may be a random vector. Let $\theta = g(\eta)$ be a (possibly vector) parameter of interest, of dimension p. Without loss we may assume that $\eta = (\theta, \lambda)$, with θ the p-dimensional interest parameter and λ a $d - p$-dimensional nuisance parameter. Suppose we wish to test a null hypothesis of the form $H_0 : \theta = \theta_0$, with θ_0 specified, or, through the familiar duality between tests of hypotheses and confidence sets, construct a confidence set for the parameter of interest θ. If $p = 1$, we may wish to allow one-sided inference, for instance a test of H_0 against a one-sided alternative of the form $\theta > \theta_0$ or $\theta < \theta_0$, or construction of a one-sided confidence limit. Let $l(\eta) = l(\eta; Y)$ be the log-likelihood for η based on Y. Also, denote by $\hat{\eta} = (\hat{\theta}, \hat{\lambda})$ the overall maximum likelihood estimator of η, and by $\hat{\lambda}_\theta$ the constrained maximum likelihood estimator of λ, for a given fixed value of θ. Inference on θ may be based on the likelihood ratio statistic, $W = w(\theta) = 2\{l(\hat{\eta}) - l(\theta, \hat{\lambda}_\theta)\}$. If $p = 1$, one-sided inference uses the signed square root likelihood ratio statistic $R = r(\theta) = \text{sgn}(\hat{\theta} - \theta)w(\theta)^{1/2}$, where $\text{sgn}(x) = -1$ if $x < 0$, $= 0$ if $x = 0$ and $= 1$ if $x > 0$. In a first-order theory of inference, the two key distributional results are that W is distributed as χ^2_p, to error of order $O(n^{-1})$, while R is distributed as $N(0, 1)$, to an error of order $O(n^{-1/2})$.

3 Exponential family and ancillary statistic models

Suppose the log-likelihood is of the form $l(\eta) = \theta s_1(Y) + \lambda^T s_2(Y) - k(\theta, \lambda) - d(Y)$, with θ scalar, so that θ is a natural parameter of a multiparameter exponential family. We wish to test $H_0 : \theta = \theta_0$ against a one-sided alternative, and do so using the signed root statistic R.

Here the conditional distribution of $s_1(Y)$ given $s_2(Y) = s_2$ depends only on θ, so that conditioning on the observed value s_2 is indicated as a means of eliminating the nuisance parameter. So, the appropriate inference on θ is based on the distribution of $s_1(Y)$, given the observed data value of s_2. This distribution is, in principle, known, since it is completely specified, once θ is fixed. In fact, this conditional inference has the unconditional (repeated sampling) optimality property of yielding a uniformly most powerful unbiased test: see, for example, Young and Smith [29, Section 7.2]. In practice however, the exact inference may be difficult to construct: the relevant conditional distribution typically requires awkward analytic calculations, numerical integrations etc., and may even be completely intractable.

In modern convention, ancillarity in the presence of nuisance parameters is generally defined in the following terms. Suppose the minimal sufficient statistic for η may be written as $(\hat{\eta}, A)$, where the statistic A has, at least approximately, a sampling distribution which does not depend on the parameter η. Then A is said to be ancillary and the conditionality principle would argue that inference should be made conditional on the observed value $A = a$.

McCullagh [22] showed that the conditional and marginal distributions of signed root statistics derived from the likelihood ratio statistic W for a vector interest parameter, but with no nuisance parameter, agree to an error of order $O(n^{-1})$, producing very similar p-values whether one conditions on an ancillary statistic or not. Severini [25] considered similar results in the context of a scalar interest parameter without nuisance parameters; see also Severini [26, Section 6.4.4]. Zaretski *et al.* [30] establish stability of the signed root statistic R, in the case of a scalar interest parameter and a general nuisance parameter. The key to their analysis is that the first two cumulants of the signed root statistic $r(\theta)$ are of the form

$$E\{r(\theta)\} = n^{-1/2} m(\eta) + O(n^{-3/2}), \quad var\{r(\theta)\} = 1 + n^{-1} v(\eta) + O(n^{-3/2}),$$

where $m(\eta)$ and $v(\eta)$ are of order $O(1)$. The third- and higher-order cumulants of $r(\theta)$ are of order $O(n^{-3/2})$; see Severini [26, Section 5.4]. This cumulant structure also holds conditionally given a statistic A assumed to be second-order ancillary; see McCullagh [22] for details of approximate ancillarity. Under conditions required for valid Edgeworth expansions, if the conditional and marginal expectations of the signed root statistic agree to an error of order $O(n^{-1})$ given the ancillary statistic A, then the conditional and marginal distributions agree to the same order of error. An intricate analysis shows that the conditional and marginal versions of $m(\eta)$ coincide, to order $O(n^{-1})$. This methodology may be readily extended to the case of a vector interest parameter θ to establish stability of signed root statistics derived from the likelihood ratio statistic W in the presence of nuisance parameters. Stability of W is immediate: the marginal and conditional distributions are both χ^2_p to error $O(n^{-1})$.

4 Analytic approximations

A detailed development of analytic methods for distributional approximation which yield higher-order accuracy in approximation of an exact conditional inference is described by Barndorff-Nielsen and Cox [2]. A sophisticated and intricate theory yields two particularly important methodological contributions. These are Bartlett corrections of the likelihood ratio statistic W and the development of analytically adjusted forms of the signed root likelihood ratio statistic R, which are specifically constructed to offer conditional validity, to high asymptotic order, in both the multiparameter exponential family and ancillary statistic contexts. Particularly central to the analytic approach to higher-order accurate conditional inference is Barndorff-Nielsen's R^* statistic (Barndorff-Nielsen, [1]).

In some generality, the expectation of $w(\theta)$ under parameter value η may be expanded as

$$\mathrm{E}_\eta\{w(\theta)\} = p\left\{1 + \frac{b(\eta)}{n} + O(n^{-2})\right\}.$$

The basis of the Bartlett correction is to modify $w(\theta)$, through a scale adjustment, to a new statistic

$$w(\theta)/\{1 + b(\eta)/n\},$$

which turns out to be distributed as χ^2_p, to an error of order $O(n^{-2})$, rather than the error $O(n^{-1})$ for the raw statistic $w(\theta)$. Remarkably, and crucially for inference in the presence of nuisance parameters, this same reduction in the order of error of an χ^2_p approximation is achievable if the scale adjustment is made using the quantity $b(\theta, \hat{\lambda}_\theta)$; see Barndorff-Nielsen and Hall [3]. Note that this result may be re-expressed as saying that the statistic

$$w^*(\theta) = \frac{p}{\mathrm{E}_{(\theta,\hat{\lambda}_\theta)}\{w(\theta)\}} w(\theta)$$

is distributed as χ^2_p to an error of order $O(n^{-2})$. Here the quantity $\mathrm{E}_{(\theta,\hat{\lambda}_\theta)}\{w(\theta)\}$ may be approximated by simulation, allowing the Bartlett correction to be carried out purely empirically, without analytic calculation.

The adjusted signed root statistic R^* has the form

$$R^* = r^*(\theta) = r(\theta) + r(\theta)^{-1}\log\{u(\theta)/r(\theta)\}.$$

Write $\eta = (\eta^1, \ldots, \eta^d)$, so that $\theta = \eta^1$ is the scalar parameter of interest, with $\lambda = (\eta^2, \ldots, \eta^d)$ a vector nuisance parameter. Let $l_{rs}(\eta) = \partial^2 l(\eta)/\partial\eta^r\eta^s$, and let $l_{\eta\eta} = (l_{rs})$ be the $d \times d$ matrix with components $l_{rs}(\eta)$ and $l_{\lambda\lambda}$ be the $(d-1) \times (d-1)$ submatrix corresponding to the nuisance parameter λ. In the exponential family context, the adjustment quantity $u(\theta)$ takes the simple form

$$u(\theta) = (\hat{\theta} - \theta)\frac{\left|-l_{\eta\eta}(\hat{\theta}, \hat{\lambda})\right|^{1/2}}{\left|-l_{\lambda\lambda}(\theta, \hat{\lambda}_\theta)\right|^{1/2}}.$$

In the ancillary statistic context the adjustment necessitates explicit specification of the ancillary statistic A and more awkward analytic calculations. For details of its construction, see Barndorff-Nielsen and Cox [2, Section 6.6].

The sampling distribution of R^* is $N(0, 1)$, to an error of order $O(n^{-3/2})$, conditionally on $A = a$, and therefore also unconditionally. Standard normal approximation to the sampling distribution of R^* therefore yields third-order (in fact, relative) conditional accuracy, in the ancillary statistic setting, and inference which respects that of exact conditional inference in the exponential family setting to the same third-order. The analytic route therefore achieves the goal of improving on the error of order $O(n^{-1/2})$ obtained from the asymptotic normal distribution of R by two orders of magnitude, $O(n^{-1})$, while respecting the conditional inference desired in the two problem classes.

5 Bootstrap approximations

The simple idea behind the bootstrap or simulation alternative to analytic methods of inference is estimation of the sampling distribution of the statistic of interest by its sampling distribution under a member of the parametric family $F(y; \eta)$, fitted to the available sample data. A recent summary of the repeated sampling properties of such schemes is given by Young [28]. We are concerned here with an analysis of the extent to which the bootstrap methods, applied unconditionally, nevertheless achieve accurate approximation to conditional inference in the exponential family and ancillary statistic settings.

DiCiccio and Young [14] show that in the exponential family context, accurate approximation to the exact conditional inference may be obtained by considering the marginal distribution of the signed root statistic R under the fitted model $F(y; (\theta, \hat{\lambda}_\theta))$, that is the model with the nuisance parameter taken as the constrained maximum likelihood estimator, for any given value of θ. This scheme yields inference agreeing with exact conditional inference to a relative error of third order, $O(n^{-3/2})$. Specifically, DiCiccio and Young [14] show that

$$\text{pr}\{R \geq r; (\theta, \hat{\lambda}_\theta)\} = \text{pr}(R \geq r | s_2(Y) = s_2; \theta)\{1 + O(n^{-3/2})\},$$

when r is of order $O(1)$. Their result is shown for both continuous and discrete models. The approach therefore has the same asymptotic properties as saddle point methods developed by Skovgaard [27] and Barndorff-Nielsen [1] and studied by Jensen [18]. DiCiccio and Young [14] demonstrate in a number of examples that this approach of estimating the marginal distribution of R gives very accurate approximations to conditional inference even in very small sample sizes: further examples are discussed in Section 6 below. A crucial point of their analysis is that the marginal estimation should fix the nuisance parameter as its constrained maximum likelihood estimator: the same third-order accuracy is not obtained by fixing the nuisance parameter at its global maximum likelihood value $\hat{\lambda}$.

Third-order accuracy can also be achieved, in principle, by estimating the marginal distributions of other asymptotically standard normal pivots, notably Wald and score statistics. However, in numerical investigations, using R is routinely shown to provide more accurate results. A major advantage of using R is its low skewness; consequently, third-order error can be achieved, although not in a relative sense, by merely correcting R for its mean and variance and using a standard normal approximation to the standardised version of R. Since it is computationally much easier to approximate the mean and variance of R by parametric bootstrapping at $(\theta, \hat{\lambda}_\theta)$ than it is to simulate the entire distribution of R, the use of mean and variance correction offers substantial computational savings, especially for constructing confidence intervals. Although these savings are at the expense of accuracy, numerical work suggests that the loss of accuracy is unacceptable only when the sample size is very small.

In theory, less conditional accuracy is seen in ancillary statistic models. Since the marginal and conditional distributions of R coincide with an error of order $O(n^{-1})$ given $A = a$, it follows that the conditional p-values obtained from R are approximated to the same order of error by the marginal p-values. Moreover, for approximating the marginal p-values, the marginal distribution of R can be approximated to an error of order $O(n^{-1})$ by means of the parametric bootstrap; the value of η used in the bootstrap can be either the overall maximum likelihood estimator, $\eta = (\hat{\theta}, \hat{\lambda})$, or the constrained maximum likelihood estimator, $\eta = (\theta, \hat{\lambda}_\theta)$. For testing the null hypothesis $H_0 : \theta = \theta_0$, the latter choice is feasible; however, for constructing confidence intervals, the choice $\eta = (\hat{\theta}, \hat{\lambda})$ is computationally less demanding. DiCiccio *et al.* [13] and Lee and Young [21] showed that the p-values obtained by using $\eta = (\theta, \hat{\lambda}_\theta)$ are marginally uniformly distributed to an error of order $O(n^{-3/2})$, while those obtained by using $\eta = (\hat{\theta}, \hat{\lambda})$ are uniformly distributed to an error of order $O(n^{-1})$ only. Numerical work indicates that using $\eta = (\theta, \hat{\lambda}_\theta)$ improves conditional accuracy as well, although, formally, there is no difference in the orders of error to which conditional p-values are approximated by using the two choices. Though in principle the order of error in approximation of exact conditional inference obtained by considering the marginal distribution of R is larger than the third-order, $O(n^{-3/2})$, error obtained by normal approximation to the sampling distribution of the adjusted signed root statistic R^*, substantial numerical evidence suggests very accurate approximations are obtained in practice. Examples are given in Section 6, and further particular examples are considered by DiCiccio *et al.* [13], Young and Smith [29, Section 11.5] and Zaretzki *et al.* [30].

In the case of a vector interest parameter θ, both the marginal and conditional distributions of $W = w(\theta)$ are chi-squared to error $O(n^{-1})$, and hence, using the χ^2_p approximation to the distribution of W achieves conditional inference to an error of second-order. Here, however, we have noted that a simple scale adjustment of the likelihood ratio statistic improves the chi-squared approximation:

$$\frac{p}{\mathrm{E}_{(\theta,\lambda)}\{w(\theta)\}} w(\theta)$$

is distributed as χ^2_p to an error of order $O(n^{-2})$. Since $\mathrm{E}_{(\theta,\lambda)}\{w(\theta)\}$ is of the form $p + O(n^{-1})$, it follows that $\mathrm{E}_{(\theta,\hat{\lambda}_\theta)}\{w(\theta)\} = \mathrm{E}_{(\theta,\lambda)}\{w(\theta)\} + O_p(n^{-3/2})$. Thus, estimation of the marginal distribution of W by bootstrapping with $\eta = (\theta, \hat{\lambda}_\theta)$ yields an approximation having an error of order $O(n^{-3/2})$; moreover, to an error of order $O(n^{-2})$, this approximation is the distribution of a scaled χ^2_p random variable with scaling factor $\mathrm{E}_{(\theta,\hat{\lambda}_\theta)}\{w(\theta)\}/p$. The result of Barndorff-Nielsen and Hall [3], that

$$\frac{p}{\mathrm{E}_{(\theta,\hat{\lambda}_\theta)}\{w(\theta)\}}w(\theta)$$

is distributed as χ^2_p to an error of order $O(n^{-2})$, shows that confidence sets constructed by using the bootstrap approximation to the marginal distribution of W have marginal coverage error of order $O(n^{-2})$.

The preceding results continue to hold under conditioning on the ancillary statistic. In particular,

$$\frac{p}{\mathrm{E}_{(\theta,\lambda)}\{w(\theta)|A=a\}}w(\theta)$$

is conditional on $A = a$, also χ^2_p to an error of order $O(n^{-2})$. The conditional distribution of W is, to an error of order $O(n^{-2})$, the distribution of a scaled χ^2_p random variable with scaling factor $\mathrm{E}_{(\theta,\lambda)}\{w(\theta)|A=a\}/p$. Generally, the difference between $\mathrm{E}_{(\theta,\lambda)}\{w(\theta)\}$ and $\mathrm{E}_{(\theta,\lambda)}\{w(\theta)|A=a\}$ is of order $O(n^{-3/2})$ given $A = a$, and using the bootstrap estimate of the marginal distribution of W approximates the conditional distribution to an error of order $O(n^{-3/2})$. Thus, confidence sets constructed from the bootstrap approximation have conditional coverage error of order $O(n^{-3/2})$, as well as marginal coverage error of order $O(n^{-2})$.

Bootstrapping the entire distribution of W at $\eta = (\theta, \hat{\lambda}_\theta)$ is computationally expensive, especially when constructing confidence sets, and two avenues for simplification are feasible. Firstly, the order of error in approximation to conditional inference remains of order $O(n^{-3/2})$ even if the marginal distribution of W is estimated by bootstrapping with $\eta = (\hat{\theta}, \hat{\lambda})$, the global maximum likelihood estimator. It is likely that using $\eta = (\theta, \hat{\lambda}_\theta)$ produces greater accuracy, however, this increase in accuracy might not be sufficient to warrant the additional computational demands. Secondly, instead of bootstrapping the entire distribution of W, the scaled chi-squared approximation could be used, with the scaling factor $\mathrm{E}_{(\theta,\hat{\lambda}_\theta)}\{w(\theta)\}/p$ being estimated by the bootstrap. This latter approach of empirical Bartlett adjustment is studied in numerical examples in Section 6. Use of the bootstrap for estimating Bartlett adjustment factors was proposed by Bickel and Ghosh [8].

6 Examples

6.1 Inverse Gaussian distribution

Let $\{Y_1, \ldots, Y_n\}$ be a random sample from the inverse Gaussian density

$$f(y;\theta,\lambda) = \sqrt{\frac{\theta}{2\pi}}\exp(\sqrt{\theta\lambda})y^{-3/2}\exp\{-\frac{1}{2}(\theta y^{-1}+\lambda y)\},\ y>0,\ \theta>0, \lambda>0.$$

The parameter of interest θ is the shape parameter of the distribution, which constitutes a two-parameter exponential family.

With $S = n^{-1}\sum_{i=1}^{n} Y_i^{-1}$ and $C = n^{-1}\sum_{i=1}^{n} Y_i$, the appropriate conditional inference is based on the conditional distribution of S, given $C = c$, the observed data value of C. This, making exact conditional inference simple in this problem, is equivalent to inference being based on the marginal distribution of $V = \sum_{i=1}^{n}(Y_i^{-1} - \overline{Y}^{-1})$. The distribution of θV is χ^2_{n-1}.

The signed root statistic $r(\theta)$ is given by $r(\theta) = \mathrm{sgn}(\hat{\theta} - \theta)\{n(\log\hat{\theta} - 1 - \log\theta + \theta/\hat{\theta})\}^{1/2}$, with the global maximum likelihood estimator $\hat{\theta}$ given by $\hat{\theta} = n/V$. The signed root statistic $r(\theta)$ is seen to be a function of V, and therefore has a sampling distribution which does not depend on the nuisance parameter λ. Since $r(\theta)$ is in fact a monotonic function of V, and the exact conditional inference is equivalent to inference based on the marginal distribution of this latter statistic, the bootstrap inference will actually replicate the exact conditional inference without error, at least in an infinite bootstrap simulation. Thus, from a conditional inference perspective, a bootstrap inference will be exact in this example.

6.2 Log-normal mean

As a second example of conditional inference in an exponential family, suppose $\{Y_1, \ldots, Y_n\}$ is a random sample from the normal distribution with mean μ and variance τ, and that we want to test the null hypothesis that $\psi \equiv \mu + \frac{1}{2}\tau = \psi_0$, with τ as nuisance parameter. This inference problem is equivalent to that around the mean of the associated log-normal distribution.

The likelihood ratio statistic is

$$w(\psi) = n[-\log\hat{\tau} - 1 + \log\hat{\tau}_0 + \frac{1}{4}\hat{\tau}_0 + \{\hat{\tau} + (\bar{Y} - \psi_0)^2\}/\hat{\tau}_0 + (\bar{Y} - \psi_0)],$$

with $\bar{Y} = n^{-1}\sum_{i=1}^{n} Y_i$, $\hat{\tau} = n^{-1}\sum_{i=1}^{n}(Y_i - \bar{Y})^2$, and where the constrained maximum likelihood estimator of the nuisance parameter τ under the hypothesis $\psi = \psi_0$ is given by $\hat{\tau}_0 = 2[\{1 + \hat{\tau} + (\bar{Y} - \psi_0)^2\}^{1/2} - 1]$.

In this example, calculation of the p-values associated with the exact conditional test is awkward, requiring numerical integration, but quite feasible: details of the test are given by Land [19]. We perform a simulation of 5000 datasets, for various sample sizes n, from the normal distribution with $\mu = 0, \tau = 1$, and consider one-sided testing of the hypothesis $H_0 : \psi = 1/2$, testing against $\psi > 1/2$. We compare the average absolute percentage relative error of different approximations to the exact conditional p-values over the 5000 replications in Table 1. Details of the methods are as follows: r is based on $N(0, 1)$ approximation to the distribution of $r(\psi)$; r^* is based on $N(0, 1)$ approximation to the distribution of $r^*(\psi)$; boot is based on bootstrap estimation of the marginal distribution of $r(\psi)$. All bootstrap results are based on 5,000,000 samples. The figures in parenthesis show the proportion of the 5000 replications where the corresponding method gave the smallest absolute percentage error. Bootstrapping the marginal distribution of the signed root statistic

Table 1. Log-normal mean problem: comparison of average absolute percentage relative errors in estimation of exact conditional p-values over 5000 replications

n	r	r^*	*boot*
5	6.718	0.476	0.367
	(0.3%)	(37.3%)	(62.4%)
10	4.527	0.154	0.136
	(0.1%)	(41.9%)	(58.0%)
15	3.750	0.085	0.077
	(0.0%)	(42.3%)	(57.7%)
20	3.184	0.054	0.050
	(0.0%)	(43.3%)	(56.7%)

is highly effective as a means of approximating the exact conditional inference for a small n, this procedure remaining competitive with the r^* approximation, which yields the same theoretical error rate, $O(n^{-3/2})$, as the sample size n increases.

6.3 Weibull distribution

As a simple illustration of an ancillary statistic model, suppose that $\{T_1, \ldots, T_n\}$ is a random sample from the Weibull density

$$f(t; \nu, \lambda) = \lambda\nu(\lambda t)^{\nu-1} \exp\{-(\lambda t)^{\nu}\}, \; t > 0, \; \nu > 0, \lambda > 0,$$

and that we are interested in inference for the parameter ν: note that $\nu = 1$ reduces to the exponential distribution. If we take $Y_i = \log T_i$, then the Y_i are an independent sample of size n from an extreme value distribution $EV(\mu, \theta)$, a location-scale family, with scale and location parameters $\theta = \nu^{-1}$, $\mu = -\log\lambda$. It is straightforward to construct exact inference for θ, conditional on the ancillary $a = (a_1, \ldots, a_n)$, with $a_i = (y_i - \hat{\mu})/\hat{\theta}$: see, for example, Pace and Salvan [24, Section 7.6].

Again, we perform a simulation of 5000 datasets, for various sample sizes n, from the Weibull density with $\nu = \lambda = 1$, and consider both one-sided and two-sided testing of the hypothesis $H_0 : \theta = 1$, in the one-sided case testing against $\theta > 1$. As before, we compare the average absolute percentage relative error of different approximations to the exact conditional p-values over the 5000 replications in Table 2. Details of the methods are as follows. For the one-sided inference: r is based on $N(0, 1)$ approximation to the distribution of $r(\theta)$; r^* is based on $N(0, 1)$ approximation to the distribution of $r^*(\theta)$; boot is based on bootstrap estimation of the marginal distribution of $r(\theta)$. For two-sided inference: w is based on χ^2_1 approximation to the distribution of $w(\theta)$; Bart is based on χ^2_1 approximation to the (empirically) Bartlett corrected $w^*(\theta)$; boot is based on bootstrap estimation of the marginal distribution of $w(\theta)$. As before, all bootstrap results are based on 5,000,000 samples, this same simulation being used for empirical Bartlett correction. Figures in parenthesis show the proportion of the 5000 replications where the corresponding method gave the smallest

Table 2. Weibull scale problem: comparison of average absolute percentage relative errors in estimation of exact conditional p-values over 5000 replications

		One-sided			*Two-sided*	
n	r	r^*	*boot*	w	*Bart*	*boot*
10	37.387	1.009	0.674	12.318	0.666	0.611
	(0.0%)	(17.1%)	(82.9%)	(0.0%)	(43.9%)	(56.1%)
20	25.473	0.388	0.397	6.118	0.185	0.227
	(0.0%)	(46.2%)	(53.8%)	(0.0%)	(63.4%)	(36.6%)
30	20.040	0.252	0.307	4.158	0.131	0.200
	(0.0%)	(60.9%)	(39.1%)	(0.0%)	(68.7%)	(31.3%)
40	17.865	0.250	0.273	3.064	0.117	0.177
	(0.0%)	(70.1%)	(29.9%)	(0.0%)	(69.7%)	(30.3%)

absolute percentage error. For both one-sided and two-sided inference, bootstrapping the marginal distribution of the appropriate statistic is highly effective for a small n, though there is some evidence that as n increases in the two-sided case, the simulation effort is better directed at estimation of the (marginal) expectation of $w(\theta)$, and the approximation to an exact conditional inference made via the chi-squared approximation to the scale-adjusted statistic $w^*(\theta)$.

6.4 Exponential regression

Our final example concerns inference on a two-dimensional interest parameter, in the presence of a scalar nuisance parameter.

Let $Y_1, \dots, Y_n$ be independent and exponentially distributed, where Y_i has mean $\lambda \exp(-\theta_1 z_i - \theta_2 x_i)$, where the z_i and x_i are covariates, with $\sum_{i=1}^n z_i = \sum_{i=1}^n x_i = 0$. The interest parameter is $\theta = (\theta_1, \theta_2)$, with λ nuisance. The log-likelihood function for (θ, λ) can be written as

$$l(\theta, \lambda) = -n \log \lambda - n\hat{\lambda}_\theta / \lambda.$$

Here $a = (a_1, \dots, a_n)$ is the appropriate conditioning ancillary statistic, with $a_i = \log y_i - \log \hat{\lambda} + \hat{\theta}_1 z_i + \hat{\theta}_2 x_i$.

The likelihood ratio statistic $w(\theta)$ is easily shown to have the simple form

$$w(\theta) = 2 \log[\frac{1}{n} \sum_{i=1}^{n} \exp\{a_i + (\theta_1 - \hat{\theta}_1) z_i + (\theta_2 - \hat{\theta}_2) x_i\}].$$

Now we perform a simulation of 2000 datasets, for various sample sizes n, from this exponential regression model with $\theta = (0, 0)$, $\lambda = 1$, and consider testing of the hypothesis $H_0 : \theta = (0, 0)$. Let $z = (54, 52, 50, 65, 52, 52, 70, 40, 36, 44, 54, 59)$ and $x = (12, 8, 7, 21, 28, 13, 13, 22, 36, 9, 87)$: these are covariate values in a lung cancer survival dataset described by Lawless [20, Table 6.3.1]. In our simulations, for

Table 3. Exponential regression problem: comparison of average absolute percentage relative errors in estimation of exact conditional p-values over 2000 replications

n	w	*Bart*	*boot*
5	12.473	1.199	0.557
	(0.4%)	(23.8%)	(75.8%)
7	5.795	1.013	0.962
	(3.5%)	(42.7%)	(53.9%)
9	4.636	0.791	0.786
	(8.8%)	(43.1%)	(48.2%)
11	4.840	1.642	1.622
	(18.4%)	(33.3%)	(48.4%)

a given n, the covariate values $(z_1, \ldots, z_n)$ and $(x_1, \ldots, x_n)$ are taken as the first n members of z and x respectively, suitable centred to have $\sum_{i=1}^n z_i = \sum_{i=1}^n x_i = 0$. Exact conditional inference for this model is detailed by Lawless [20, Section 6.3.2]. Exact conditional p-values based on the likelihood ratio statistic W are, following Barndorff-Nielsen and Cox [2, Section 6.5], obtained by numerical integration of the exact conditional density of $\hat{\theta}$ given a , as described by Lawless [20, Section 6.3.2], over the appropriate set of values of $\hat{\theta}$ which give a value of W exceeding the observed value.

The average absolute percentage relative error of different approximations to the exact conditional p-values over the 2000 replications are given in Table 3. Now w is based on χ^2_2 approximation to the distribution of $w(\theta)$; Bart is based on χ^2_2 approximation to the (empirically) Bartlett corrected $w^*(\theta)$; boot is based on bootstrap estimation of the marginal distribution of $w(\theta)$. As before, all bootstrap results are based on 5,000,000 samples, this same simulation being used for empirical Bartlett correction, and the figures in parenthesis show the proportion of the 2000 replications where the corresponding method gave the smallest absolute percentage error. Now, exact conditional p-values appear to be effectively approximated by the marginal distribution of the likelihood ratio statistic, though the empirical Bartlett correction is quite comparable.

7 Conclusions

Marginal simulation approaches to approximation of an exact conditional inference have been shown to be highly effective, in both multiparameter exponential family and ancillary statistic models.

For inference on a scalar natural parameter in an exponential family, the appropriate exact one-sided conditional inference can be approximated to a high level of accuracy by marginal simulation of the signed root likelihood ratio statistic R. This procedure considers the sampling distribution of R under the model in which the interest parameter is fixed at its null hypothesis value and the nuisance parameter is specified as its constrained maximum likelihood value, for that fixed value of the interest parameter. The theoretical rate of error in approximation of the exact conditional inference is the same ($O(n^{-3/2})$) as that obtained by normal approximation to the distribution of the adjusted signed root statistic R^*, and excellent approximation is seen with small sample sizes. Similar practical effectiveness is seen with small sample sizes n in ancillary statistic models, though here the theoretical error rate of the marginal simulation approach, $O(n^{-1})$, is inferior to that of the analytic approach based on R^*.

In ancillary statistic models, where interest is in a vector parameter, or in two-sided inference, based on the likelihood ratio statistic W, on a scalar interest parameter, two marginal simulation approaches compete. The first uses the simulation to approximate directly the sampling distribution of W and the second approximates the marginal expectation of W, this then being the basis of empirical Bartlett correction of W. The two methods are seen to perform rather similarly in practice, with direct approximation of the distribution of the stable statistic W yielding particularly good results in small sample size situations.

References

1. Barndorff-Nielsen, O.E.: Inference on full or partial parameters based on the standardized signed log likelihood ratio. Biometrika **73**, 307–22 (1986)
2. Barndorff-Nielsen, O.E., Cox, D.R.: Inference and Asymptotics. Chapman & Hall, London (1994)
3. Barndorff-Nielsen, O.E., Hall, P.G.: On the level-error after Bartlett adjustment of the likelihood ratio statistic. Biometrika **75**, 374–8 (1988)
4. Basu, D.: On statistics independent of a complete sufficient statistic. Sankhya **15**, 377–80 (1955)
5. Basu, D.: The family of ancillary statistics. Sankhya **21**, 247–56 (1959)
6. Basu, D.: Problems relating to the existence of maximal and minimal elements in some families of statistics (subfields). *Proc. Fifth Berk. Symp. Math. Statist. Probab.* **1**. University of California Press, Berkeley CA, 41–50 (1965)
7. Basu, D.: On the elimination of nuisance parameters. J. Amer. Statist. Assoc. **72**, 355–66 (1977)
8. Bickel, P.J., Ghosh, J.K.: A decomposition for the likelihood ratio statistic and the Bartlett correction – a Bayesian argument. Ann. Statist. **18**, 1070–90 (1990)
9. Birnbaum, A.: On the foundations of statistical inference (with discussion). J. Amer. Statist. Assoc. **57**, 269–306 (1962)
10. Brazzale, A.R., Davison, A.C., Reid, N.: Applied Asymptotics: Case Studies in Small-Sample Statistics. Cambridge University Press, Cambridge (2007)
11. Cox, D.R.: Some problems connected with statistical inference. Ann. Math. Stat. **29**, 357–72 (1958)
12. Cox, D.R.: Local ancillarity. Biometrika **67**, 279–86 (1980)
13. DiCiccio, T.J., Martin, M.A., Stern, S.E.: Simple and accurate one-sided inference from signed roots of likelihood ratios. Can. J. Statist. **29**, 67–76 (2001)
14. DiCiccio, T.J., Young, G.A.: Conditional properties of unconditional parametric bootstrap procedures for inference in exponential families. Biometrika **95**, 747–58 (2008)

15. Efron, B., Hinkley, D.V.: Assessing the accuracy of the maximum likelihood estimator: Observed versus expected Fisher information (with discussion). Biometrika **65**, 457–87 (1978)
16. Fisher, R.A.: Two new properties of mathematical likelihood. Proc. R. Soc. Lond. A **144**, 285–307 (1934)
17. Fisher, R.A.: The logic of inductive inference. J.R. Statist. Soc. **98**, 39–82 (1935)
18. Jensen, J.L.: The modified signed likelihood statistic and saddlepoint approximations. Biometrika **79**, 693–703 (1992)
19. Land, C.E.: Confidence intervals for linear functions of normal mean and variance. Ann. Statist. **42**, 1187–205 (1971)
20. Lawless, J.: Statistical Models and Methods for Lifetime Data. Wiley, New York (1982)
21. Lee, S.M.S., Young, G.A.: Parametric bootstrapping with nuisance parameters. Stat. Prob. Letters **71**, 143–53 (2005)
22. McCullagh, P.: Local sufficiency. Biometrika **71**, 233–44 (1984)
23. McCullagh, P.: Conditional inference and Cauchy models. Biometrika **79**, 247–59 (1992)
24. Pace, L., Salvan, A.: Principles of Statistical Inference: from a Neo-Fisherian Perspective. World Scientific Publishing, Singapore (1997)
25. Severini, T.A.: Conditional properties of likelihood-based significance tests. Biometrika **77**, 343–52 (1990)
26. Severini, T.A.: Likelihood methods in Statistics. Oxford University Press, Oxford (2000)
27. Skovgaard, I.M.: Saddlepoint expansions for conditional distributions. J. Appl. Probab. **24**, 875–87 (1987)
28. Young, G.A.: Routes to higher-order accuracy in parametric inference. Aust. N.Z. J. Stat. **51**, 115–26 (2009)
29. Young, G.A., Smith, R.L.: Essentials of Statistical Inference. Cambridge University Press, Cambridge (2005)
30. Zaretzki, R., DiCiccio, T.J., Young, G.A.: Stability of the signed root likelihood ratio statistic in the presence of nuisance parameters. Submitted for publication (2010)

Monte Carlo simulation methods for reliability estimation and failure prognostics

Enrico Zio

Abstract. Monte Carlo Simulation (MCS) offers a powerful means for modeling the stochastic failure behaviour of engineered structures, systems and components (SSC). This paper summarises current work on advanced MCS methods for reliability estimation and failure prognostics.

Key words: Monte Carlo simulation, reliability, subset sampling, line sampling, failure prognostics, particle filtering

With respect to the estimation of the reliability of SSC by Monte Carlo simulation, the challenge is due to the small failure probabilities involved which may bring a significant computational burden. The recently developed Subset Simulation (SS) and Line Sampling (LS) methods are here illustrated and shown capable of improving the MCS efficiency in the estimation of small failure probabilities. The SS method is founded on the idea that a small failure probability can be expressed as a product of larger conditional probabilities of some intermediate events: with a proper choice of intermediate events, the conditional probabilities can be made sufficiently large enough to allow accurate estimation with a small number of samples. The LS method employs lines instead of random points in order to probe the failure domain of interest. An "important direction" is determined, which points towards the failure domain of interest; the high-dimensional reliability problem is then reduced to a number of conditional one-dimensional problems which are solved along the "important direction".

With respect to failure prognosis, in general terms the primary goal is to indicate whether the SSC of interest can perform its function throughout its lifetime with reasonable assurance and, in the case it cannot, to estimate its Time To Failure (TTF), i.e. the lifetime remaining before it can no longer perform its function. The soundest model-based approaches to the state estimation of a SSC build a posterior distribution of its unknown state by combining the distribution assigned a priori with the likelihood of the observations of measurements of parameters or variables related to the SSC state. In this Bayesian setting, a Monte Carlo simulation method which is becoming popular is the so-called particle filtering method which approximates the state distributions of interest by discrete sets of weighed 'particles' representing random

Mantovan, P., Secchi, P. (Eds.): Complex Data Modeling and Computationally Intensive Statistical Methods

trajectories of system evolution in the state space, and whose weights are estimates of the probabilities of the trajectories. The potential of this method is here illustrated.

1 Introduction

MCS is a powerful method for modeling and evaluating the stochastic failure behaviour of a complex SSC, thanks to its flexibility and to the indifference of the model solution to its complexity and dimensionality. The method is based on the repeated sampling of realisations of SSC configurations and evolutions.

In mathematical terms, the probability of SSC failure can be expressed as a multi-dimensional integral of the form

$$P(F) = P(\underline{x} \in F) = \int I_F(\underline{x}) q(\underline{x}) d\underline{x}, \tag{1}$$

where $\underline{x} = \{x_1, x_2, \ldots, x_j, \ldots, x_n\} \in \Re^n$ is the vector of the random states of the components, i.e. the random configuration of the system, with multi-dimensional probability density function (PDF) $q : \Re^n \to [0, \infty)$, $F \subset \Re^n$ is the failure region and $I_F : \Re^n \to \{0, 1\}$ is an indicator function such that $I_F(\underline{x}) = 1$, if $\underline{x} \in F$ and $I_F(\underline{x}) = 0$, otherwise.

In practical cases, the multi-dimensional integral (1) cannot be easily evaluated by analytical methods nor by numerical schemes. On the other hand, MCS offers an effective means for estimating the integral, because the method does not suffer from the complexity and dimension of the domain of integration, albeit it implies the non-trivial task of sampling from the multi-dimensional probability density function (PDF). Indeed, the MCS solution to (1) entails that a large number of samples of the values of the component states vector be drawn from $q(\cdot)$; an unbiased and consistent estimate of the failure probability is then simply computed as the fraction of the number of samples that lead to failure. However, a large number of samples (inversely proportional to the failure probability) is necessary to achieve an acceptable estimation accuracy: in terms of the integral in (1) this can be seen as due to the high dimensionality n of the problem and the large dimension of the relative sample space compared to the failure region of interest [27].

In this respect, advanced simulation approaches which overcome some of the limitations of IS are offered by Subset Simulation (SS) [3] and Line Sampling (LS) [17]. These MCS schemes are here presented by showing their application to a structural reliability model of literature, i.e., the Paris-Erdogan thermal fatigue crack growth model [22]. The problem is rather challenging as it entails estimating failure probabilities of the order of 10^{-7}.

Failure prognosis is becoming a more and more attractive and challenging task in Reliability, Availability, Maintainability and Safety (RAMS). The attractiveness of prognostics comes from the fact that by predicting the evolution of the SSC state, it is possible to provide advanced warning and lead time for preparing the necessary corrective actions to maintain the SSC in safe and productive operation.

However, in real SSC, often the states cannot be directly observed; on the other hand, measurements of parameters or variables related to the SSC states are available, albeit usually affected by noise and disturbances. Then, the problem becomes that of inferring the SSC state from the measured parameters. Two general approaches exist: i) the model-based techniques, which make use of a quantitative analytical model of SSC behaviour [31] and ii) the knowledge-based or model-free methods, which rely on empirical models built on available data of SSC behaviour [19,26].

The soundest model-based approaches to the SSC state estimation problem build a posterior distribution of the unknown states by combining the distribution assigned a priori with the likelihood of the observations of the measurements actually collected [7,8]. In this Bayesian setting, the estimation method most frequently used in practice is the Kalman filter, which is optimal for linear state space models and independent, additive Gaussian noises. In this case, the posterior distributions are also Gaussian and can be computed exactly, without approximations.

In practice the dynamic evolution of many SSC is non-linear and the associated noises are non-Gaussian [16]. For these cases, approximate methods, e.g. analytical approximations of extended Kalman (EKF) and Gaussian-sum filters and numerical approximations of the grid-based filters [1] can be used, usually at large computational expenses. Alternatively, one may resort to MCS methods also known as *particle filtering* methods, which are capable of approximating the continuous and discrete distributions of interest by a discrete set of weighed 'particles' representing random trajectories of system evolution in the state space, and whose weights are estimates of the probabilities of the trajectories [6,9]. As the number of samples becomes large, the Monte Carlo approximation yields a posterior pdf representation which is equivalent to its functional description and the particle filter approaches the optimal Bayesian TTF prediction.

In this paper, particle filtering is presented by showing its application to the Paris-Erdogan thermal fatigue crack growth model [22].

2 The subset and line sampling methods for realiability estimation

An advanced MCS approach to SSC reliability estimation problems characterised by low-failure probabilities is the SS, originally developed to tackle the multi-dimensional problems of structural reliability [3]. In this approach, the failure probability is expressed as a product of conditional failure probabilities of some chosen intermediate events, whose evaluation is obtained by simulation of more frequent events. The problem of evaluating small failure probabilities in the original probability space is thus replaced by a sequence of simulations of more frequent events in the conditional probability spaces. The necessary conditional samples are generated through successive Markov Chain Monte Carlo (MCMC) simulations [12, 15, 20], gradually populating the intermediate conditional failure regions until the final target failure region is reached.

A recent alternative approach to the solution of the rare-event reliability estimation problems is the Line Sampling (LS) technique, also originally developed to efficiently tackle the multi-dimensional problems of structural reliability [17]. *Lines*, instead of random *points*, are used to probe the failure domain of the high-dimensional problem under analysis [23]. An "important direction" is optimally determined to point towards the failure domain of interest and a number of conditional, one-dimensional problems are solved along such direction, in place of the high-dimensional problem [23]. The approach has been shown to always perform better than standard MCS; furthermore, if the boundaries of the failure domain of interest are not too rough (i.e., almost linear) and the "important direction" is almost perpendicular to them, the variance of the failure probability estimator could be ideally reduced to zero [17].

As an example of application let us consider the thermal fatigue crack growth model based on the deterministic Paris-Erdogan model which describes the propagation of a manufacturing defect due to thermal fatigue [22]. The evolution of the size x of a crack defect satisfies the following equation:

$$\frac{dx}{dN} = C \cdot (f(R) \cdot \Delta K)^n, \tag{2}$$

where N is the number of fatigue cycles, C and n are parameters depending on the properties of the material, $f(R)$ is a correction factor which is a function of the material resistance R, and ΔK is the variation of the intensity factor, defined as

$$\Delta K = \Delta s \cdot Y(x) \cdot \sqrt{\pi x}. \tag{3}$$

In (25), Δs is the variation of the uniform loading (stress) applied to the system and $Y(x)$ is the shape factor of the defect. Let $S_i = \Delta s_i$ be the variation of the uniform normal stress at cycle $i = 1, 2, \ldots, N$. The integration of equation (2) gives

$$\int_{x_0}^{x_N} \frac{dx}{(Y(x)\sqrt{\pi x})^n} = C \cdot \sum_{i=1}^{N} (f(R) \cdot S_i)^n, \tag{4}$$

where x_0 and x_N are the initial and final size of the crack defect, respectively. In (4) the following approximation can be adopted:

$$\sum_{i=1}^{N} = (f(R) \cdot S_i)^n \approx (T - T_0) \cdot N \cdot (f(R) \cdot S)^n, \tag{5}$$

where T and T_0 are the initial and final times of the thermal fatigue treatment (of N cycles).

The system is considered failed when the size x_N of the defect at the end of the N cycles exceeds a critical dimension x_c, i.e.:

$$x_c - x_N \leq 0, \tag{6}$$

which in the integral form (4) reads

$$\psi(x_c) - \psi(x_N) \leq 0, \tag{7}$$

where

$$\psi(x) = \int_{x_0}^{x} \frac{dx'}{(Y(x') \cdot \sqrt{\pi x'})^n}. \tag{8}$$

Using (27), a safety margin $M(T)$ can then be defined as follows:

$$M(T) = \int_{x_0}^{x_c} \frac{dx}{(Y(x) \cdot \sqrt{\pi x})^n} - C \cdot (T - T_0) \cdot N \cdot (f(R) \cdot S)^n. \tag{9}$$

The failure criterion can then be expressed in terms of the safety margin (9):

$$M(T) \leq 0. \tag{10}$$

The probabilistic representation of the uncertainties affecting the nine variables x_0, x_c, T_0, T, C, n, $f(R)$, N and S (hereafter named $x_1, x_2, x_3, x_4, x_5, x_6, x_7, x_8$ and x_9, respectively), leads to the following definition of the probability of system failure $P(F)$:

$$\begin{aligned} P(F) &= P[M(T) \leq 0] \\ &= P\left[\int_{x_0}^{x_c} \frac{dx}{(Y(x) \cdot \sqrt{\pi x})^n} - C \cdot (T - T_0) \cdot N \cdot (f(R) \cdot S)^n \leq 0\right], \end{aligned} \tag{11}$$

or

$$\begin{aligned} P(F) &= P[M(T) \leq 0] \\ &= P\left[\int_{x_2}^{x_1} \frac{dx}{(Y(x) \cdot \sqrt{\pi x})^{x_6}} - x_5 \cdot (x_4 - x_3) \cdot x_8 \cdot (x_7 \cdot x_9)^{x_6} \leq 0\right]. \end{aligned} \tag{12}$$

It is worth noting the highly non-linear nature of expressions (11) and (12) which increases the complexity of the problem.

The characteristics of the PDFs of the uncertain variables are summarised in Table 1; the value of the exact (i.e., analytically computed) failure probability, $P(F)$ is also reported in the last row of Table 1.

For fair comparison, in the application, all simulation methods have been run with the same total number of samples ($N_T = 40{,}000$). The efficiency of the methods has been evaluated in terms of the Figure Of Merit (FOM) defined as $1/(\hat{\sigma}^2 t_{comp})$, where t_{comp} is the computational time required by the simulation method (Table 2). It can be seen that SS performs consistently better than standard MCS. On the other hand, LS outperforms SS, in spite of the fact that the determination of the sampling important direction $\underline{\alpha}$ and the calculations of the conditional one-dimensional failure probability estimates require much more than N_T system analyses by the model; this is due to the accelerated convergence rate that can be attained by the LS method with respect to SS.

Finally, both the SS and LS methods have been shown to perform well, and in fact outperform other conventional sampling methods in the reliability assessment of a variety of systems of different complexity, e.g. structures [17, 24, 28, 29, 32], multi-state and continuous-state series-parallel systems [33], thermal-hydraulic passive systems [34, 35], etc.

Table 1. Probability distributions and parameters (i.e., means and standard deviations) of the uncertain variables $x_1, x_2, \ldots, x_9$ of the thermal fatigue crack growth model; the last row reports the value of the corresponding exact (i.e., analytically computed) failure probability, $P(F)$ [13, 14]. *Exp* = exponential distribution; *LG* = Lognormal distribution; *N* = Normal distribution

x_1 (x)	$Exp(0.81{\cdot}10^{-3})$
x_2 (x_c)	$N(21.4{\cdot}10^{-3}, 0.214{\cdot}10^{-3})$
x_3 (T)	0
x_4 (T)	40
x_5 (C)	$LG(1.00{\cdot}10^{-12}, 5.75{\cdot}10^{-13})$
x_6 (n)	3.4
x_7 $(f(R))$	2
x_8 (N)	$N(20, 2)$
x_9 (S)	$LG(200, 20)$
$P(F)$	$1.780{\cdot}10^{-5}$

Table 2. Results of the application of standard MCS, SS and LS to the reliability estimation of the thermal fatigue crack growth model

	$\hat{P}(F)$	$\hat{\sigma}$	N_{sys}	*FOM*
Standard MCS	$1.780{\cdot}10^{-5}$	$2.269{\cdot}10^{-5}$	40000	$4.860{\cdot}10^{4}$
SS	$1.130{\cdot}10^{-5}$	$1.653{\cdot}10^{-6}$	39183	$9.341{\cdot}10^{6}$
LS	$1.810{\cdot}10^{-5}$	$2.945{\cdot}10^{-8}$	81999	$1.188{\cdot}10^{13}$

3 Particle filtering for failure prognosis

In general terms, the problem of estimating the states of a dynamical system is usually carried out in a discrete time domain by considering a dynamical system for which both a set of measurements and a theoretical model linking the system states among themselves and with the measurements, are available.

In correspondence of a sequence of equidistant discrete times t, where t stands for $\tau_t = t \cdot \Delta t$, $(t = 0, 1, 2, \ldots)$, it is desired to infer the unknown (hidden) state $x_t \equiv x(\tau_t)$ on the basis of all the previously estimated state values $x_{0:t-1} \equiv (x_0, x_1, \ldots, x_{t-1})$ and of all the measurements $z_{0:t} \equiv (z_0, z_1, \ldots, z_t)$, $z_i \equiv z(\tau_i)$, collected up to time t by a set of sensors. Both the system states and the measurements, which may be multi-dimensional variables, are affected by inherent noises.

In a Bayesian context the associated filtering problem amounts to evaluating the posterior distribution $p(x_t|z_{0:t})$. This can be done by sampling a large number N_s of time sequences $\{x^i_{0:t}\}_{i=1}^{N_s}$ from a suitably introduced importance function $q(x_{0:t}|z_{0:t})$ [8]. In the state space, this sample of sequences represents an ensemble of trajectories of state evolution similar to those simulated in particle transport phenomena: the problem is then that of utilising the ensemble of N_s simulated trajectories for filtering out the unobserved trajectory of the real process.

The filtering distribution $p(x_t|z_{0:t})$ is the marginal of the probability $p(x_{0:t}|z_{0:t})$, i.e. the multiple integral of this latter with respect to $x_{t_0}, x_{t_1}, \ldots, x_{t-1}$ in $[-\infty, \infty]^t$ *viz.*, $p(x_t|z_{0:t}) = \int p(x_{0:t}|z_{0:t})dx_{0:t-1}$. The integration may be formally extended to include also the variable x_t by means of a δ-function, i.e. $p(x_t|z_{0:t}) = \int p(x_{0:t-1}, u|z_{0:t})\delta(x_t - u)dx_{0:t-1}du$. By resorting to the importance sampling method previously recalled, a large number N_s of trajectories $\{x^i_{0:t}\}^{N_s}_{i=1}$ is first sampled from the importance function $q(x_{0:t}|z_{0:t})$ and then the integral is approximated as

$$\begin{aligned} p(x_t|z_{0:t}) &= \int \left[\frac{p(x_{0:t-1}, u|z_{0:t})}{q(x_{0:t-1}, u|z_{0:t})}\delta(x_t - u)\right] q(x_{0:t-1}, u|z_{0:t})dx_{0:t-1}du \\ &\approx \sum_{i=1}^{N_s} w^i_t \delta(x_t - x^i_t), \end{aligned} \tag{13}$$

where the weights w^i_t of the estimation are

$$w^i_t = \frac{p(x^i_{0:t}|z_{0:t})}{q(x^i_{0:t}|z_{0:t})}, \tag{14}$$

which can be recursively computed as

$$w^i_t = w^i_{t-1}\frac{p(z_t|x^i_t)p(x^i_t|x^i_{t-1})}{q(x^i_t|x^i_{t-1})}. \tag{15}$$

Unfortunately, the trajectories sampled according to the procedure illustrated suffer from the so-called *degeneracy phenomenon*: after few samplings, most of the N_s weights in (15) become negligible so that the corresponding trajectories do not contribute to the estimate of the PDF of interest [8].

A possible remedy to this problem is to resort to the so-called resampling method [2], based on the bootstrap technique which essentially consists of sampling balls from an urn with replacement [10, 11]. At each time *t*, N_s samplings with replacement are effectuated from an urn containing N_s balls; the *i-th* ball is labelled with the pair of known numbers $\{w^i_t, x^i_t\}$ and it will be sampled with a probability proportional to the weight value w^i_t; a record of the sampled pairs is maintained; at the end of these *N* multinomial samplings, there is a good chance that the recorded sample will contain several replicas of the balls with larger weights (in other words, that the final record will contain several identical copies of the same label), whereas a corresponding number of balls with smaller weights will not appear in the sample (in other words, a corresponding number of labels is lost from the sample).

In the described bootstrap procedure, it is evident that the sampled weights are i.i.d. so that the same weight *1/N*$_s$ may be assigned to all sampled pairs. Then, the filtering procedure continues with the original pairs $\{w^i_t, x^i_t\}^{N_s}_{i=1}$ replaced by new pairs $\left\{1/N_s, x^{i^*}_t\right\}^{N_s}_{i^*=1}$ in which several i^* may correspond to the same i in the original pairs. Equation (13) then becomes

$$p(x_t|z_{0:t}) \approx \sum_{i=1}^{N_s} \frac{p(z_t|x_t^i)p(x_t^i|x_{t-1}^i)}{q(x_t^i|x_{t-1}^i)} \delta(x_t - x_t^i) \approx \sum_{i=1}^{N_s} \frac{1}{N_s} \delta(x_t - x_t^{i^*}). \tag{16}$$

A pseudo-code describing the basic steps of the procedure is:

- at *t=0*, a sequence $\{x_0^i\}_{i=1}^{N_s}$ is sampled from *p*(x_0);
- at the generic time *t*>*0*:
  - a value z_t is measured (or simulated if we are dealing with a case study);
  - a sequence $\{x_t^i\}_{i=1}^{N_s}$ is sampled from the given $q(x|x_{t-1}^i)$;
  - the N_s likelihoods $\{p(z_t|x_t^i)\}_{i=1}^{N_s}$ are evaluated;
  - the weights w_t^i required by the described resampling procedure are evaluated from (15) in which $w_{t-1}^i = 1$;
  - the resampling procedure is performed and the obtained $x_t^{i^*}$ yield the resampled realisations of the states at time *t;*
  - the $x_t^{i^*}$-range, $X_t = \max_{i^*}(x_t^{i^*}) - \min_{i^*}(x_t^{i^*})$, is divided in a given number of intervals and the mean probability values in these intervals are given by the histogram of the $x_t^{i^*}$.

Let us consider the Paris-Erdogan model introduced in the previous section; for simplification, but with no loss of generality, it is assumed that $f(R) = 1$ and ΔK is directly proportional to the square root of x [25]:

$$\Delta K = \beta\sqrt{x}, \tag{17}$$

where β is a constant which may be determined from experimental data.

In this case, the intrinsic stochasticity of the process is inserted in the model as follows [25]:

$$\frac{dx}{dN} = e^{\omega} C(\beta\sqrt{x})^n, \tag{18}$$

where $\omega \sim N(0, \sigma_\omega^2)$ is a white Gaussian noise. For ΔN sufficiently small, the state-space model (18) can be discretised to give:

$$x_t = x_{t-1} + e^{\omega_t} C(\Delta K)^n \Delta N, \tag{19}$$

which represents a non-linear Markov process with independent, non-stationary degradation increments.

The degradation state x_t is generally not directly measurable. In the case of non-destructive ultrasonic inspections a logit model for the observation z_t can be introduced [30]:

$$\ln\frac{z_t}{d - z_t} = \beta_0 + \beta_1 \ln\frac{x_t}{d - x_t} + \upsilon_t, \tag{20}$$

where d is the component material thickness, $\beta_0 \in (-\infty, \infty)$ and $\beta_1 > 0$ are parameters to be estimated from experimental data and υ is a white Gaussian noise such that $\upsilon \sim N(0, \sigma_\upsilon^2)$.

Introducing the following standard transformations,

$$y_t = \ln \frac{z_t}{d - z_t}, \tag{21}$$

$$\mu_k = \beta_0 + \beta_1 \ln \frac{x_t}{d - x_t}, \tag{22}$$

then, $Y_t \sim N(\mu_t, \sigma_v^2)$ is a Gaussian random variable with conditional cumulative distribution function (cdf):

$$F_{Y_t}(y_t \,|x_t) = P(Y_t < y_t \,|x_t) = \Phi\left(\frac{y_t - \mu_t}{\sigma_v}\right), \tag{23}$$

where $\Phi(u)$ is the cdf of the standard normal distribution $N(0, 1)$.

The conditional cdf of the measurement z_t related to the degradation state x_t is then:

$$F_{Z_t}(z_t \,|x_t) = F_{Y_t}\left(\ln \frac{z_t}{d - z_t} \,|x_t\right) = \Phi\left(\frac{1}{\sigma_v}\left(\ln \frac{z_t}{d - z_t} - \mu_t\right)\right), \tag{24}$$

with corresponding probability density function (pdf):

$$f_{Z_t}(z_t \,|x_t) = \frac{1}{\sqrt{2\pi}\,\sigma_v} e^{-\frac{1}{2}\left(\frac{\ln \frac{z_t}{d-z_t} - \mu_t}{\sigma_v}\right)^2} \frac{d}{z_t(d - z_t)}. \tag{25}$$

The particle filtering estimation method has been applied to a case study of literature in which the parameters of the state equation (19) are $C = 0.005$, $n = 1.3$ and $\beta = 1$, whereas those in the measurement equation (20) are $\beta_0 = 0.06$, and $\beta_1 = 1.25$. The process and measurement noise variances are $\sigma_\omega^2 = 2.89$ and $\sigma_v^2 = 0.22$, respectively. The cycle time step is $\Delta N = 1$. The component is assumed failed when the crack depth $x \geq x_c = d = 100$, in arbitrary units. The five measurements at time steps N_1, N_2, N_3, N_4, N_5 are generated from (20), based on the crack growth dynamics simulated according to (19). The values of the measurements are $z_1 = 0$, $z_2 = 0.8$, $z_3 = 2.9$, $z_4 = 17.9$ and $z_5 = 47.0$. In Figure 1, the true degradation state (dots) is compared to the particle filter estimate (dotted line): the performance of the particle filter is satisfactory, with a Root Mean Squared Error $RMSE = 7.9$ and a coverage, $cov = 72\%$ as defined in [5]. The CPU-time per time step of evolution is $4.6 \cdot 10^{-1}$ s. Furthermore, the posterior probability density function of the component's remaining lifetime τ_t at the inspection time t can be estimated by properly modifying the particle filter algorithm described above. In particular, the estimate can be expressed as:

$$\hat{p}(\tau_t \,|z_{0:t}) = \sum_{i=1}^{N_s} w_t^i \delta\left(\tau_t - t^i\right), \tag{26}$$

where t^i is the time step at which the component failure occurs in the *i-th* simulated crack growth trajectory.

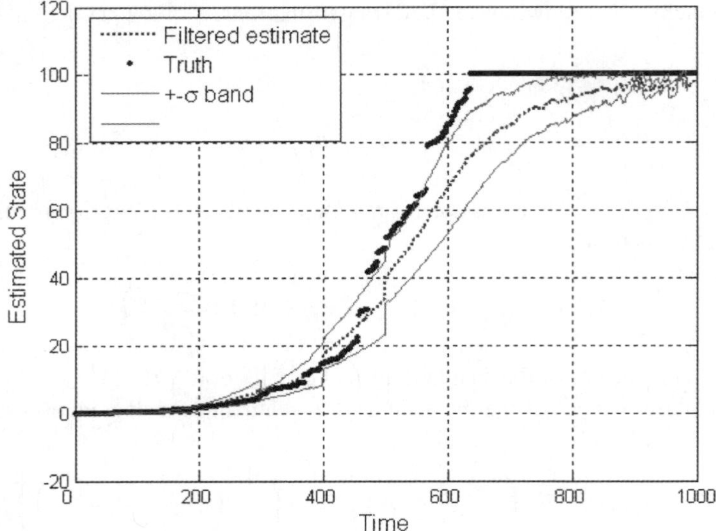

Fig. 1. Comparison of the true time evolution of the crack depth (dots) with the particle filter-estimated mean of the posterior distribution (dotted line), with $\pm 1\sigma_t$ bands (solid): five measurements $z_1 = 0$, $z_2 = 0.8$, $z_3 = 2.9$, $z_4 = 17.9$ and $z_5 = 47.0$ taken at times $N_1 = 100$, $N_2 = 200$, $N_3 = 300$, $N_4 = 400$, $N_5 = 500$

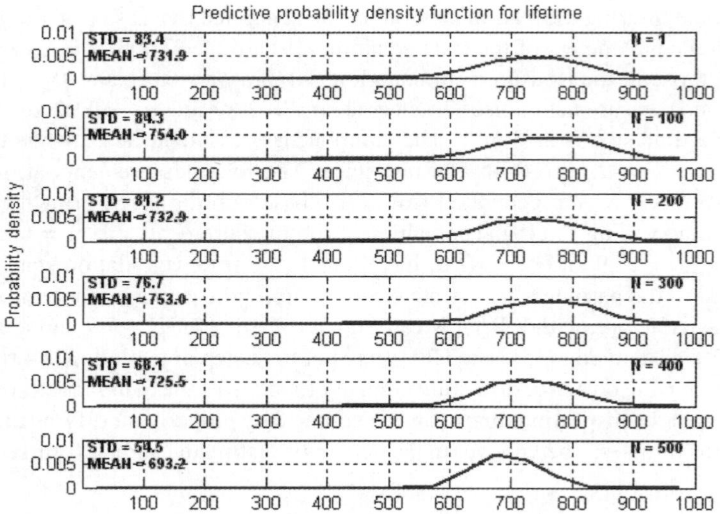

Fig. 2. Lifetime posterior probability density functions: five measurements $z_1 = 0$, $z_2 = 0.8$, $z_3 = 2.9$, $z_4 = 17.9$ and $z_5 = 47.0$ taken at times $N_1 = 100$, $N_2 = 200$, $N_3 = 300$, $N_4 = 400$, $N_5 = 500$ [5]

Operatively, the algorithm for the remaining lifetime distribution estimation can be divided into two parts:

- Off-line simulation of crack growth trajectories:
  - N_s particles are simulated off-line, before starting to observe the component's degradation in real-time;
  - when a particle reaches the full material thickness $x_c = d$, component failure occurs and the corresponding failure time t^i is collected.
- On-line estimation of the distribution of the remaining lifetime:
  - at the inspection times t_k= $1, N_1, N_2, N_3, N_4$ and N_5, the particles' weights w_t^i are updated on the basis of the last available measurement;
  - the lifetime distribution at time t can be estimated by constructing a histogram of the N_s failure times t^i weighed on the corresponding values ofw_t^i.

The uncertainty in the lifetime distribution predictions (Figure 2) decreases as the measurements become available.

Finally, particle filtering has been shown effective in a number of diagnostic and prognostic applications [21], including hybrid systems [4, 18].

4 Conclusions

The rapid increase in computing power has rendered, and will continue to render, more and more feasible the use of Monte Carlo methods for engineering calculations. In the past, restrictive modeling assumptions had to be introduced to fit the models to the numerical methods available for their solution, at the cost of moving away from reality and at the risk of obtaining sometimes dangerously misleading results. Thanks to the inherent flexibility of Monte Carlo simulation, these assumptions can be relaxed, so that realistic operating rules can be accounted for in the models for RAMS applications.

This paper has illustrated recent developments in the Monte Carlo simulation method with respect to its use for the estimation of the reliability and the failure prognostics of SSC. With the aid of examples, the potential associated with the advanced techniques of SS, LS and particle filtering have been demonstrated. SS and LS offer clever ways out of the rare event problem which affects reliability estimation in practice. In SS the problem is tackled by breaking the small failure probability evaluation task into a sequence of estimations of larger conditional probabilities. During the simulation, more frequent samples conditional to intermediate regions are generated from properly designed Markov chains. The strength of the method lies in the generality of its formulation and the straightforward algorithmic scheme. The method has been proven more effective than standard MCS. The LS method employs *lines* instead of random *points* in order to probe the high-dimensional failure domain of interest. An "important direction" is optimally determined to point towards the failure domain of interest and a number of conditional, one-dimensional problems are solved along such direction, in place of the original high-dimensional problem. In the case the boundaries of the failure domain of interest are not too rough (i.e., approximately linear) and the "important direction" is almost perpendicular to them, only few simulations suffice to arrive at a failure probability with acceptable confidence. Of particular advantage of Line Sampling is its robustness: in the worst possible case where the "important direction" is selected orthogonal to the (ideal) optimal direction, line sampling performs at least as well as standard Monte Carlo simulation.

The particle filtering method seems to offer significant potential for successful application in failure prognostics, since it is capable of handling non-linear dynamics and of dealing with non-Gaussian noises at no further computational or model design expenses. As such, it represents a valuable prognostic tool which can drive effective condition-based maintenance strategies for improving the availability, safety and cost effectiveness of plant operation.

References

1. Anderson, B.D., Moore, J.B.: Optimal Filtering. Prentice Hall, Englewood Cliffs, NJ (1979)
2. Arulampalam, M.S., Maskell, S., Gordon, N., Clapp, T.: A Tutorial on Particle Filters for Online Nonlinear/Non-Gaussian Bayesian Tracking. IEEE Trans. On Signal Proc. **50**(2), 174–188 (2002)
3. Au, S.K., Beck, J.L.: Estimation of small failure probabilities in high dimensions by subset simulation. Probabilistic Engineering Mechanics **16**(4), 263–277 (2001)
4. Cadini, F., Peloni, G., Zio, E.: Detection of the Time of Failure of a Hybrid System by Particle Filtering in a Log-Likelihood Ratio Approach, accepted for publication. In: *Proceedings of IEEE-Prognostics & System Health Management Conference, 12–14 January 2010, Macau* (2010)
5. Cadini, F., Zio, E., Arram, D.: Model-Based Monte Carlo State Estimation for Condition-Based Component Replacement. Reliability Engineering and System **94**(3), 752–758 (2008)
6. Djuric, P.M., Kotecha, J.H., Zhang, J., Huang, Y., Ghirmai, T., Bugallo, M.F., Miguez, J.: Particle Filtering. IEEE Signal Processing Magazine, 19–37 (2003)
7. Doucet, A.: On Sequential Simulation-Based Methods for Bayesian Filtering, Technical Report. University of Cambridge, Dept. of Engineering, CUED-F-ENG-TR310 (1998)
8. Doucet, A., De Freitas, J.F.G., Gordon, N.J.: An Introduction to Sequential Monte Carlo Methods. In: Doucet, A., de Freitas, J.F.G., Gordon, N.J. (eds.) *Sequential Monte Carlo in Practice*. Springer-Verlag, New York (2001)
9. Doucet, A., Godsill, S., Andreu, C.: On Sequential Monte Carlo Sampling Methods for Bayesian Filtering. Statistics and Computing **10**, 197–208 (2000)
10. Efron, B.: Bootstrap Methods: Another Look at the Jacknife. Annals of Statistics **7**, 1–26 (1979)
11. Efron, B., Tibshirani, R.J.: An Introduction to the Bootstrap. Chapman and Hall, New York (1993)
12. Fishman, G.S.: Monte Carlo: concepts, algorithms, and applications. Springer, New York (1996)
13. Gille, A.: Evaluation of failure probabilities in structural reliability with Monte Carlo methods. ESREL '98, Throndheim (1998)

14. Gille, A.: Probabilistic numerical methods used in the applications of the structural reliability domain. PhD Thesis, Universitè Paris 6 (1999)
15. Hastings, W.K.: Monte Carlo sampling methods using Markov chains and their applications. Biometrika **57**, 97–109 (1970)
16. Kitagawa, G.: Non-Gaussian State-Space Modeling of Nonstationary Time Series. Journal of the American Statistical Association **82**, 1032–1063 (1987)
17. Koutsourelakis, P.S., Pradlwarter, H.J., Schueller, G.I.: Reliability of structures in high dimensions, Part I: algorithms and application. Probabilistic Engineering Mechanics **19**, 409–417 (2004)
18. Koutsoukos, X., Kurien, J., Zhao, F.: Monitoring and diagnosis of hybrid systems using particle filtering methods. In: *Proceedings of the Fifteenth International Symposium on the Mathematical Theory of Networks and Systems (MTNS), University of Notre Dame, Notre Dame, USA* (2002)
19. Marseguerra, M., Zio, E., Baraldi, P., Popescu, I.C., Ulmeanu, P.: A Fuzzy Logic-based Model for the Classification of Faults in the Pump Seals of the Primary Heat Transport System of a Candu 6 Reactor. Nuclear Science and Engineering **153**(2), 157–171 (2006)
20. Metropolis, N., Rosenbluth, A.W., Rosenbluth, M.N., Taller, A. H.: Equations of state calculations by fast computing machines. Journal of Chemical Physics **21**(6), 1087–1092 (1953)
21. Orchard, M.E., Vachtsevanos, G.: A particle filtering framework for failure prognosis. In: *Proceedings of WTC2005, World Tribology Congress III, Washington D.C., USA* (2005)
22. Paris, P.C.: A rational analytic theory of fatigue. The trend of engineering at the university of Washington **13**(1), 9 (1961)
23. Pradlwarter, H.J., Pellissetti, M.F., Schenk, C.A., Schueller, G.I., Kreis, A., Fransen, S., Calvi, A., Klein, M.: Computer Methods in Applied Mechanics and Engineering **194**, 1597–1617 (2005)
24. Pradlwarter, H.J., Schueller, G.I., Koutsourelakis, P.S., Charmpis, D.C.: Application of line sampling simulation method to reliability benchmark problems. Structural Safety **29**, 208–221 (2007)
25. Provan, J.W. (ed.): Probabilistic fracture mechanics and reliability. Martinus Nijhoff Publishers, Dordrecht (1987)
26. Reifman, J.: Survey of Artificial Intelligence Methods for Detection and Identification of Component Faults in Nuclear Power Plants. Nucl. Technol. **119**(76), (1997)
27. Schueller, G.I.: On the treatment of uncertainties in structural mechanics and analysis. Computers and Structures **85**, 235–243, (2007)
28. Schueller, G.I., Pradlwarter, H.J.: Benchmark study on reliability estimation in higher dimensions of structural systems – An overview. Structural Safety **29**, 167–182 (2007)
29. Schueller, G.I., Pradlwarter, H.J., Koutsourelakis, P.S.: A critical appraisal of reliability estimation procedures for high dimensions. Probabilistic Engineering Mechanics **19**, 463–474 (2004)
30. Simola, K., Pulkkinen, U.: Models for non-destructive inspection data. Reliability Engineering and System Safety **60**, 1–12 (1998)
31. Willsky, A.S.: A survey of design methods for failure detection in dynamic systems. Automatica **12**, 601–611 (1976)
32. Zio, E., Pedroni, N.: Reliability estimation by advanced Monte Carlo simulation. In: *Simulation Methods for Reliability and Availability of Complex Systems*. Springer, London 3–40 (2010)
33. Zio, E., Pedroni, N.: Reliability analysis of discrete multi-state systems by means of subset simulation. *ESREL 2008, European Safety and Reliability Conference*, September 22–25, 2008, Valencia, Spain, 709–716 (2008)

34. Zio, E., Pedroni, N.: Estimation of the functional failure probability of a thermal-hydraulic passive system by subset simulation. Nuclear Engineering and Design **239**(3), 580–599 (2009)
35. Zio, E., Pedroni, N.: Functional failure analysis of a thermal-hydraulic passive system by means of line sampling. Reliability Engineering and System Safety **9**(11), 1764–1781 (2009)

Printed in July 2010